The Ecosystems Revolution

Mark Everard

The Ecosystems Revolution

Mark Everard
University of the West of England (UWE),
Bristol
United Kingdom

ISBN 978-3-319-81089-8   ISBN 978-3-319-31658-1   (eBook)
DOI 10.1007/978-3-319-31658-1

Printed on acid-free paper

This Palgrave Macmillan imprint is published by Springer Nature
The registered company is Springer International Publishing AG Switzerland

ALSO BY MARK EVERARD

- Breathing Space: The Natural and Unnatural History of Air (2015)
- The Hydropolitics of Dams: Engineering or Ecosystems? (2013)
- Common Ground: The Sharing of Land and Landscapes for Sustainability (2011)
- The Business of Biodiversity (2009)
- PVC: Reaching for Sustainability (2008)
- Water Meadows: Living Treasures in the English Landscape (2005)

CONTENTS

Contents

LIST OF FIGURES

CHAPTER 1

Introduction

Keywords Revolution • Symbiotic • Sustainable development • Interdependence • Breakthroughs • Symbiocene • Decision-making

The Ecosystems Revolution: Co-creating a Symbiotic Future is all about humanity's relationship with the natural world, how it has shifted throughout our evolutionary journey, and how we urgently need to accelerate its evolution on a far more symbiotic basis. The book draws upon and integrates a number of themes—natural and artificial selection processes in evolution and decision-making, how revolutions are constructed and perceived, directed versus random change, and the history and necessary future trajectory of the human story—seeking guidance on achievement of a sustainable future secured by a symbiotic relationship with the ecosystems that constitute its vital underpinnings.

Chapter 2, 'Of this Earth', considers the integrally co-evolved and interdependent nature of all life, from microbes to humans and the workings of the entire biosphere, highlighting the indivisibility of all human activities from the rest of nature. This interdependence underlies today's diverse and pressing sustainability challenges, including both their causes and their potential solutions. This recognition illuminates the need for an 'ecosystems revolution', progressively repositioning the workings of nature's supportive processes into governance systems to build a future of greater security, wellbeing and opportunity.

Chapter 3, 'Breakthroughs in the ascent of humanity', plots the trajectory of human development through the lens of the materials and

© The Editor(s) (if applicable) and The Author(s) 2016
M. Everard, *The Ecosystems Revolution*,
DOI 10.1007/978-3-319-31658-1_1

technologies we have harnessed to further our own prospects. These have been characterised as a series of so-called 'revolutions' in the manipulation of natural resources. A generally narrow focus on immediate advantages accruing from largely fortuitous 'evolutionary' innovations has frequently also generated multiple unintended consequences, emphasising the need for greater cognisance of systemic ramifications for people and supporting ecosystems as a basis for the next societal revolution.

Chapter 4, 'Chance or choice?', reviews the nature of natural selection, a primary concept in the theory of evolution, including the application of selection principles to the evolution of ideas, technologies and products. It contrasts the multi-factorial nature of natural selection with the often narrow framing of artificial selection, which externalises many of the impacts of human innovations on ecosystems. These impacts, in turn compromise the capacities of affected ecosystems to sustain human needs, as a form of natural selection process. This highlights the need for a new type of revolution in human development that is directed rather than relying on fortuitous innovations, and is also guided by a broader framework of 'artificial selection' principles more closely aligned to the complexity of the natural world. It also challenges current conceptions of sustainable development that implicitly assume stationarity when, in fact, the ongoing pace of ecosystem decline and the burgeoning of the human population require us to raise our vision to one that encompasses the progressive rebuilding of ecosystem capacity and resilience.

Chapter 5, 'Reanimating the landscape', draws upon a range of inspiring community-based projects across the developing world where restoration of degraded landscapes has regenerated ecosystems and human livelihoods in a positively reinforcing cycle. Parallels are drawn with emerging approaches to restoration of catchments and their functions for pollution control, water resource protection, flood management, and other outcomes on an increasingly integrated, nature-based way. Examples are drawn from across the world where ecosystem restoration is protecting and increasing human security, economic benefits and opportunity, highlighting the importance of investment in the natural infrastructure vital for securing human wellbeing into the future. However, the difficulties of navigating a transition to a broader, systemic paradigm are significant, threatening as this broadening of conception may appear to mind sets shaped by currently established norms and vested interests founded on more reductive perspectives.

Chapter 6, 'A revolutionary journey', explores how an ecosystems revolution is already under way, as evidenced by incremental modifications to the broader formal and informal policy environment of the developed world over the past century and more. The dependencies and impacts of major policy areas on ecosystems and their services are reviewed through selected examples, emphasising the need for far greater internalisation of the benefits and vulnerabilities of supporting ecosystems, integrated across policy spheres and societal sectors, if continuing human opportunity is to be secured.

Chapter 7, 'Co-creating the Symbiocene', recognises that human pressures will continue to exert significant influence on global ecosystems, whether we chose to direct ourselves on a progressive path or permit continuing decline through inaction. What is undoubtedly required, if sustainability becomes our guiding principle, is to achieve increasing symbiosis between natural processes, with their associated 'natural selection' forces, and the choices and 'artificial selection' criteria that humanity applies to direct it towards that chosen future. This directed revolution to achieve symbiosis between the natural processes that shaped the Holocene with the human pressures that currently, in their unreconstructed state, shape the Anthropocene, would constitute a new synergistic and sustainable epoch: the Symbiocene. A framework for decision-making is presented, backed up by a range of worked examples across policy areas, before concluding with thoughts on the unique influences we all bring to bear through our day-to-day choices and actions, all of which influence, to unpredictable degrees, the kind of future we are co-creating.

The Ecosystems Revolution: Co-creating a Symbiotic Future is packed with practical and positive examples, inspiring us that, for all the attendant negative trends, a revolution is possible. This will not be a revolution brought about by force or violence; rather, it is one that we will co-create, indeed are co-creating, through shared understanding, aspiration, and consideration of the ramifications of our incremental decisions and actions. It is about empowerment and engagement in a journey, for it is not the ecosystems that require a revolution; they have always and will always adapt and survive. It is about us co-creating a revolution that progressively embeds the multiple values and importance of thriving, regenerating ecosystems into the ways that we think, act and live lives of potentially expanding opportunity and fulfilment.

CHAPTER 2

Of This Earth

Abstract 'Of this Earth' considers the integrally co-evolved and interdependent nature of all life, from microbes to humans and the workings of the entire biosphere, highlighting the indivisibility of all human activities from the rest of nature. This interdependence underlies today's diverse and pressing sustainability challenges, including both their causes and their potential solutions. This recognition illuminates the need for an 'ecosystems revolution', progressively repositioning the workings of nature's supportive processes into governance systems to build a future of greater security, wellbeing, and opportunity.

Keywords Cycles • Ecosystems • Biosphere • Indivisibility • Technology • Natural limits

The natural world of dynamically interactive solid matter, water, gases, electromagnetic fields, and profuse living organisms upon, within, and with which we live is a sphere. This is true physically, but it is also true chemically and ecologically. There are no sharp corners where things are not joined up as endless cycles. The sun that our home planet orbits radiates energy that plant life captures to fuse simple components into complex organic matter through photosynthetic processes, powering efficient and seamless biospheric cycles of carbon, phosphorus, nitrogen, oxygen, and other substances, tuned through 3.85 billion years of evolution.

© The Editor(s) (if applicable) and The Author(s) 2016
M. Everard, *The Ecosystems Revolution*,
DOI 10.1007/978-3-319-31658-1_2

There have been many blind evolutionary alleys, the price of which has been extinction. There have been catastrophes too, including meteoric collisions such as the best-known one that ended the age of the dinosaurs. Amongst other catastrophes is the evolution of photosynthetic processes some 2.5 billion years ago that, whilst expunging virtually all pre-existing life forms evolved in the absence of highly reactive atmospheric oxygen, culminated in more energetic and diverse ecosystems powered by respiratory oxidation of organic matter. It also enabled life to populate shallow waters and land surfaces, shielded from destructive wavelengths of radiation from the sun by stratospheric ozone formed from those raised atmospheric oxygen levels.

Gaia Theory likens the tight co-evolution and co-dependence of life forms operating homeostatically as a contiguous whole to a form of super-organism within which each living component not only interacts intimately with all others, but does so in ways that contribute to conditions favourable for maintaining the endless planetary cycles and processes essential for the protection and sustenance of all life.[1]

MICROBIAL PLANET

We know more about, and have invested substantially more in exploration of, the surface of the moon nearly a quarter-million miles distant across the vacuum of space than the ocean's abyss of our home planet. But even these dark depths, with their crushing pressures of up to 1100 atmospheres (at the bottom of the 10,994 metres deep Mariana Trench) and temperatures at or below 0 °C, are well understood compared to the ecology of the most fundamental life support systems sustaining our health, prospects for wealth creation, quality of life, and future resilience.

We have, at least in the more accessible and populated parts of the world, charted much of the macroscopic flora and fauna responsible for a great deal of primary production, herbivory, carnivory, and remobilisation of energy and matter from dead organisms. As they are well documented, we will not relate what is already known about the contribution of the diversity of more conspicuous life forms to planetary cycles. By contrast, the microbial 'foot soldiers' responsible for the bulk of nature's great biogeochemical cycles remain barely recognised, characterised, and understood. This hidden treasure of life, obscured from our gaze by its microscopic dimensions, is not merely vital but also staggeringly diverse and of great cumulative mass. So it is worth spending a little time getting to know it, all the better to understand the complexity and integrated nature of the ecosystems of our home planet.

It is to this largely unrecognised and unloved microbial bestiary that we owe virtually everything. Some of its constituents, including various prokaryotic (lacking a discrete nucleus) bacteria and archaea, barely differ from what we believe to be the first life forms to evolve on Earth some 3.85 billion years ago. Yet still they, and the complex processes they perform as tightly co-evolved players with countless viruses, fungi, protozoa, algae, and diverse other life forms far smaller than the acuity of the human eye, remain essential cogs in the great planetary machinery upon which all of life not only depends but from which it arose.

Bacteria may typically only be around 1 μm (micrometre) in size, some longer but not wider. However, a teaspoon of productive soil generally contains between 100 million and 1 billion bacteria, the equivalent in biomass to two cows per acre.[2] These soil bacteria perform a wide range of important roles, simplistically categorised into four functional groups. First of all, most bacteria are decomposers, consuming organic compounds such as metabolic waste and dead matter, releasing nutrients and converting energy into forms useful to other organisms. A second group of bacteria, referred to as mutualists, form partnerships with plants. These include nitrogen-fixing bacteria that associate with the roots of some plants, benefitting from the habitat that the plants provide and some substances which they release into the soil and, in exchange, converting inert atmospheric nitrogen into forms available to the host plant. A third group of bacteria comprises pathogens that make their living by attacking other organisms, and include a number of disease-causing microorganisms. A fourth group, known as lithotrophs or chemoautotrophs, obtain their energy by metabolising chemical compounds other than those based on carbon, including, for example, nitrogen, sulphur, iron, or hydrogen, and so contribute directly to the recycling and bioavailability of these elements. The collective actions of these soil bacteria have major implications for waste breakdown including pollutant decontamination, the cycling of matter and energy, the movement of water through soils, soil structure, as well as wider services including control of plant and other diseases. In soils, the greatest concentrations of bacteria are close to the root systems of plants, known as the rhizosphere. Bacteria also pervade all terrestrial and aquatic habitats as well as being present in the atmosphere. Across these media, they perform a bewildering diversity of functions, unseen, yet vital to the fundamental cycles of nature.

Other vital, yet substantially underappreciated constituents of the complex microbial ecology of our world are the fungi, a hugely varied group of organisms comprising around 1.5 million (estimates ranging from 0.7 to 5 million) species globally. This is more than six times as diverse as all groups of flowering plants combined. Fungi do not build organic matter like plants, but release enzymes outside of their bodies, breaking down matter to release constituent substances that can then be absorbed and metabolised. They comprise a diverse range of organisms from single cellular forms, including yeasts such as what we use in brewing and baking, through to better-known larger organisms built from masses of tiny filaments including edible fungi such as mushrooms and some types of toadstools. The fungi have varying medicinal and other applications, as well as being agents of disease such as crop-destroying rusts and smuts. Since 1969, fungi have been recognised as belonging to their own kingdom of organisms discrete from plants and animals. Many fungi form intimate relationships with plants that range from the harmful to the beneficial and the symbiotic. Amongst these symbionts are the 'ectomycorrhiza'—around 5000 species of fungi that form sheaths around the root tips of approximately 10% of known plant species—the fungi receiving sugars from the plant in return for greatly enhancing the plant's ability to take up water and nutrients from the soil. It is thought that root-associated fungi enabled the initial colonisation of land by plants nearly 600 million years ago. Other benefits to host plants include protection against herbivores and resistance to toxins and pathogens. Fungi are also the main decomposers of organic material in soils and many other ecosystems, including breaking down wood, dead animals and plants, excreted substances, and other matter, releasing and recycling constituent chemicals and energy. Other fungi produce biologically active substances, some of which, including antibiotics and fermented products such as alcohol, have been exploited by humans. The huge diversity of forms and life styles of fungi means that they are adapted to all global ecosystems from the poles to jungles and deserts and as internal constituents of many organisms including humans. In many ways, fungi constitute vital connectors, recyclers, and active agents in nearly all terrestrial and aquatic ecosystems and, despite their general invisibility both to our eyes and in the ways we use and manage the natural world, they are integral agents in the fundamental processes that sustain life on this planet. Despite their extraordinary diversity and substantial ecological and economic importance, providing an essential and irreplaceable service to all planetary life by recycling nutrients, fungi remain vastly under-studied and underappreciated in comparison to visible plants and animals.

The archaea were initially classified as bacteria under the name archaebacteria, but are now classified as a discrete kingdom of single-celled, prokaryotic microorganisms. Their diversity and functions remain poorly understood, although archaeal biochemistry, including the constituents of their cell membranes, is unique. Archaea also exploit a greater diversity of energy sources than eukaryotic organisms (cells with nuclei and organelles surrounded by discrete membranes), including not only organic compounds but also a wider range including ammonia, metal ions, and hydrogen gas, with some also capable of using sunlight as an energy source. Archaea occur in a diversity of environments including as extremophiles (organisms adapted to living in harsh conditions) found in environments such as hot springs, ocean floor geothermal vents, and salt lakes. Although amongst the smallest living organisms, archaea in oceanic plankton may be one of the most abundant groups of organisms on the planet. Collectively, archaea play significant roles in carbon, nitrogen, and other vital natural cycles.

The term *protozoa* is regarded today as outmoded by modern taxonomic understanding, an all-embracing term covering an estimated 30,000 species spanning discrete groups of organisms sharing the similar attributes of being mostly (but not exclusively) unicellular and eukaryotic. Many are motile (capable of independent motion) using flagellae (long fine hairs), cilia (multiple fine hair), or pseudopodia (foot-like extensions from soft cell edges). Protozoans are limited to moist or aquatic habitats, although this can include fine moist surface films in and upon soils and biological matter. Many protozoan species are symbionts, whilst others are parasites or predators on other microorganisms. Some protozoans absorb food through their cell membranes whilst others engulf fine particulate matter, and they include predators upon unicellular or filamentous algae, bacteria, and microfungi, playing a significant role in controlling their populations. Some are parasites, such as species of *Plasmodium* (the causative agent of malaria), trypanosomes, and other disease-causing agents in humans and other animals and plants. The protozoans are important elements of the microfauna of soils, aquatic systems, and many other environments, some stimulating decomposition and others digesting cellulose in the rumen of cows and the guts of termites. Protozoans constitute important elements of food webs, playing important albeit frequently overlooked roles in nutrient mobilisation.

The algae too are a diverse and important group of eukaryotic, mainly photosynthetic organisms ranging in size from the microscopic and single-celled to larger multicellular 'water plants' such as the Charaphytes and giant kelps, some species of which may be up to 50 metres long despite lacking the vascular system that characterises most groups of higher plants. Not all algae are photosynthetic, some breaking down organic matter to derive energy. However, the vast majority perform photosynthesis. The net contribution of algae to oxygen generation and primary production of organic matter is both substantial and significant; oceanic phytoplankton, comprising mainly microscopic algae suspended in seawater, produce somewhere between 50% and 85% of the oxygen content in the planet's air.[3]

The interactions of microorganisms in ecosystems are close and of profound significance, comprising tightly co-evolved relationships that play fundamental roles in the biospheric cycling of matter and energy. These microbial interactions extend to larger organisms, including some noted above: as symbiotic ectomycorrhiza on the roots of higher plants, parasites, disease-causing organisms, decomposers, but also as internal constituents of larger organisms. These include, as we have seen, protozoan digestion of cellulose in the rumen of cows and the guts of termites, and so forth. The microbial flora of the human gut collectively acts as an additional essential 'organ', breaking down food and aiding digestion as well as providing nourishment, regulating epithelial development, and contributing to immunity. The human microbiome—the aggregate of microorganisms residing on the surface and in deep layers of skin, in the saliva, and oral mucosa, and in the conjunctiva, reproductive and gastrointestinal tracts—includes bacteria, fungi, and archaea which, according to a study,[4] outnumber human cells by a factor of 10 to 1. The human microbiome is fundamentally important, affecting many dimensions of health and behaviour. However, as for the microbiomes of all other larger organisms, and indeed the workings of whole ecosystems from the minute to the whole biosphere, it remains substantially under-researched and hence underappreciated.

Interactions between microscopic and other organisms are, however, far more profound than even this. The origins of organelles (subcellular functional structures) in eukaryotic cells are believed to have been through the symbiotic inclusion of microbes into host cells. Many algae, for example, have primary chloroplasts (photosynthetic organelles) derived from endosymbiotic cyanobacteria and other smaller algae. Under the endosymbiotic theory, it is believed that several other key organelles within eukaryotic cells originated as symbioses between separate single-celled organisms. This explains, for example, why these organelles are commonly surrounded by double rather than single membranes and, like mitochondria (organelles that break down sugars to provide energy for other cellular processes), have their own discrete heritable DNA.

A wide range of other organisms invisible to the human eye due to their microscopic size include groups of animals such as rotifers, nematodes, tardigrades, and flukes. And, of course, there is the cumulate metabolic activity of all of the visible and massive plants and animals with which we are more familiar. But the key point for our current analysis is how integrated and instrumental all of life is—and particularly the microscopic life that we largely overlook—to the sustainable cycles of planet Earth.

HUMAN PLANET

Having said that we will pay less attention to the larger, better-known flora and fauna of this planet whose contributions to its cycles are more widely known, there is one species to which we will devote disproportionate attention: humans. This is because, owing to our evolutionary gifts, we are, as far as we know, the only species likely to read this book, interpret characters on the page, and process the information they convey. The collective metabolism of our activities is as much interdependent with the biosphere as that of any other species, be it minuscule or gigantic, and as subject to the same natural laws. The fact that we have not considered and arranged our lives in the light of this basic biophysical reality lies at the root of the serious problems now confronting humanity.

Humanity arose integrally within the tightly coherent whole of the biosphere. For all our qualitative differences—greater degrees of consciousness and foresight, innovative and co-learning capacities, and a range of other features not forgetting our infamous opposable thumbs—we remain as co-dependent with the biosphere that spawned us as the bee or ant is to the colony it serves. Indeed, humanity is not merely indivisible from, but evolved as a wholly owned subsidiary of nature, our endobiomes determining our overall health just as much as our dependence on external living systems for breathing, drinking, eating, excreting, and stimulation. Yet, although inseparably creatures of the planet's endless biospheric cycles, we uniquely have created sharp edges and protruding corners in this natural sphere. We humans have done so through depletion of resources, disruption of natural cycles, and also the creation of a uniquely human phenomenon—wastes of all kinds—alien to the cyclic, re-assimilative workings of nature.

Yet, for all our transgressions, we remain intimately plumbed into the energetic and material flows of the biosphere. Consequently, all we do is shaped by and in turn shapes the character and supportive properties of the living whole. Some may argue therefore that all that humans do is inherently 'natural', and that the ways in which nature adapts to the pressures we place upon it are part of some evolutionary plan. However, the compelling evidence is that evolution has no such preconceived plan, but is a reaction to chance mutations that may better fit some to greater survival prospects whilst condemning others to eventual extinction. Further compelling evidence—including the current catastrophic loss of species and their contributions to ecosystem functioning and resilience, of declines in soil quantity and quality, of nutrient enrichment and climate change wrought by remobilisation of substances sequestered into rock away from the biosphere over evolutionary timescales—emphasises that unbridled technological innovation and resource appropriation has not been achieved without adverse consequences. Our development path may have enabled human numbers to explode, and material expectations of quality of life to have risen for a global minority, but it has been consequent from mining the very natural capital that underwrites a secure future for all in the long term and the prospects for many in the present.

The so-called undeveloped societies—those living most directly resource-dependent lifestyles—often remain in close synergy with the primary water, soil, atmospheric, and biological resources that support them, often with sophisticated formal and informal governance arrangements to assure sustainable and equitable resource use and sharing. The 'developed' world, on the other hand, increasingly appropriates natural flows of water and productivity, and emits wastes to all environmental media at a scale that not only exceeds natural regenerative capacities but also now jeopardises the very foundations of long-term wellbeing. It does so at a magnitude that threatens all of humanity, and the vitality and balance of all planetary life.

Worse still, we in the already developed world create expectations and promote pathways of economic and industrial-scale development to the developing world that are made in our own unsustainable image, intensifying collective jeopardy.

REALISING OUR INDIVISIBILITY WITHIN NATURE

All ecosystems are intimately interlinked as contiguous wholes, at all scales from the microscopic to the planet's interactions with energy flows from our home star and the rhythmic tidal pull of the moon.

As my three prior books in this series—*Common Ground*,[5] *The Hydropolitics of Dams*,[6] and *Breathing Space*[7]—address at some length, land and landscapes, the water cycle, and the air/atmosphere system constitute vital ecosystems. However, these principal environmental media are distinct only in so far as we choose to classify them in such reductive terms. In reality, all form indivisible elements of an internally interactive biospheric whole in which life is a key agent of exchange. As just one example, up to 93% of the dry mass of a mature oak tree, a mighty and sturdy solid object, is captured from tiny gaseous carbon dioxide molecules drawn from the surrounding air, melded with water and nutrients gathered also from the air or drawn up from the soil and fused through of solar radiation by the alchemy of photosynthesis.[8] Oxygen excreted as a waste from photosynthetic processes is captured by other organisms as a crucial input to respiratory functions, and also contributes to the ozone layer in the lower stratosphere that protects terrestrial life from harmful radiation from space. Microbes around the tree's roots play vital roles in remobilising the nutrients used by the tree from soil minerals, and falling leaves build soil structure and support a diversity of co-evolved organisms from the microscopic to the largest herbivores, which collectively play important roles in the transfer of water, chemical substances, and energy across landscapes within which trees are integral.

We and the sum total of our activities are entirely subsidiary to this vitally interdependent biosphere, which not only supports and shapes humanity but also is reciprocally shaped by the ways in which we use, abuse, and manage it. Even at our most technically sophisticated, for example in space travel, we take our home ecosystems with us whether as stored resources or by modelling them through technological trickery that recycles essential biospheric resources such as air and water. As discussed in Breathing Space,'*No man is an island*'[9] given the intimacy of feedback mechanisms, not merely with others of our species but with all dimensions of planetary life and environmental media. Despite the human gifts of imagination and innovation, all we do and can be rests on, and as intimately affects, the ecosystems that produced and constantly support us. We are, in an unbreakable sense, immersed in nature all our days, and from our genesis through to whatever kind of future we permit it to continue to provide for us.

As also described in detail in my prior books spanning the three principal environmental media—and which therefore will not be repeated here—so many of the sustainability challenges that we face today stem in one way or another from overlooking the inevitable implications of our actions on nature's productive, supporting, enriching, and regulatory processes. By overlooking the finite assimilative capacities of air at a local scale, for example, we create urban health issues related to the build-up of pollutants of various kinds. At the same time, overloading the air system at regional scale gives rise to wider-scale and transboundary issues such as acid rain whilst, at global scale, the cavalier disposal of greenhouse and ozone-depleting gases threatens wholesale changes in the climate as well as reductions in the stratospheric shield against damaging radiation from space. Equally, we contaminate water bodies with nutrient substances, organic matter, exotic chemicals, and metals with a host of deleterious effects, also abstracting and diverting water flows with implications for geological stability, ecosystem support, and soil productivity. The world also appears to be moving into a sixth mass extinction. This sweeping claim is substantiated by an authoritative study from the USA,[10] which clearly demonstrates that current extinction rates for mammal and other vertebrate species over the last century are over one hundred times greater than 'background' rates. More than 400 vertebrate species have been lost since 1900, a scale of loss that would normally be seen over a period of up to 10,000 years. Whilst it is still possible to avert a dramatic decay of biodiversity and the subsequent loss of ecosystem services, the paper concludes, the window of opportunity for change is rapidly closing.

Destruction of nature—from forest cover and soil extent and quality to marine fish stocks, wetlands, and coastal mangroves—undermines a host of natural processes and services of vital yet often formerly overlooked benefit to human health, wealth creation, and quality of life, destabilising resilience evolved into ecosystems over billions of years and so increasing their and our vulnerability to future pressures. There are many examples from across the world and throughout human history in which over-exploitation and eventual overwhelming of nature's supportive capacities, exacerbated by competition for scarce and dwindling resources, have lain at the root of the rise and fall of civilisations that may have once seemed invincible.[11]

The twin questions facing humanity at this point in history are simply, *'Why are we facing such a constellation of pressures threatening our own long-term viability?'* and *'Do we have the foresight and courage to make proportionate and concerted change?'* In short, do we wish to secure a future with a reasonable expectation of achieving our potential, or will our inherited short-term greed condemn us to eking out a living on, and increasingly competing for, a declining resource of damaged and degrading supportive ecosystems?

We have already crossed a Rubicon of global capacity, the increasingly resource-hungry lifestyles of the planet's burgeoning human population depleting supportive ecosystems at a manifestly unsustainable pace that presages increasing limitations to future progress and more immediate triggers for conflict. Many, for example, now recognise that the sheer scale of impacts stemming from human activities have marked a transition from the Holocene, an epoch dating back some 11,700 years that has been defined by natural forces since the end of the last major 'ice age' of the Pleistocene, propelling us into a new geological age defined as the Anthropocene in which humans are becoming the dominant influence on Earth's ecosystems.[12] Undoubtedly, our recent historic trajectory has been naïve about its wider systemic ramifications. But, knowing what we now know, we can claim neither that we lack sufficient understanding, nor that ignorance of natural law can absolve us of guilt and consequence from our actions and our continuing failures to act.

Substantial revision of human lifestyles is essential if we are to secure a decent future, a global revolution that has at its core recognition and integration within the profound and irreplaceable values of ecosystems, their processes, and beneficial services. We are an ingenious, learning species, attributes that have stimulated prior revolutions in water management, agriculture, industry, tool innovation and use, exploitation of materials and other species of plants, microbes, and animals; in the innovation of weapons, medicine, information, and communications technology; and in so many other ways. As the saying goes, the Stone Age did not end because we ran out of stones; rather, we discovered or innovated something better. Formerly, we may have defined 'better' in terms of extension of technological reach, improved shelter, and security of food supply. But today's called-for revolution—looking at the diversity of sustainability challenges created by the short-termism of recent historic technical innovation—differs in scale and character.

The revolution that we now need is a more conscious quest to optimise our collective prospects for living secure and fulfilled lives into the future through an increasingly symbiotic relationship with the ecosystems essential for supporting our needs into the future. It is also a revolution that will best insulate our economic activities and lifestyles from 'shocks' arising from unforeseen factors such as resource scarcities, collapsing natural processes, and extreme weather. This journey towards a symbiotic vision is, in addition, one that will reward sustainable innovation and create a more secure collective future. It is necessarily a revolution that must be more deliberate, collaborative, and far-sighted. Above all, it is one that once again recognises that, for all our technological prowess and sense of emancipation, we remain as resource-dependent as the tribal farmer, as the salmon, and as the hedge sparrow. Technology, after all, merely serves to extend our access to nature's beneficence, as for example the plumbing of clean water into cities and homes and of wastewater out of them to be decontaminated by technologically accelerated microbial breakdown processes, or access to food produced from remote soils and oceans. Beneath all of this sophistication is nature, underpinning the needs of rich and poor, urban and rural alike. The necessary revolution is one that recognises and integrates into societal governance systems fully our total and continuing dependence upon nature, the most fundamental resource base that we are today depleting at such an alarming rate through cultural short-termism, with a dire prognosis for the wellbeing of all.

The ecosystems revolution is a necessary rediscovery of our biospheric roots. However, this is no regressive journey. It is not, as some caricature it, about ditching the comforts and trappings of modern life to return to living in caves and foraging directly from the land. Humanity has no such reverse gear, seemingly hard-wired to innovate and move forwards. To date, 'forwards' has been measured by immediate, often competitive advantage, and certainly rewarded by another of our creations—the market economy—on that basis. Tomorrow, 'forwards' has to become judged by broader, more enlightened criteria that include some form of feedback from the future: How will this innovation make use of natural resources sympathetically with their regenerative capacities, what are the consequences for all, and can further innovations around this idea better secure benefits for humanity that enhance our collective long-term wellbeing?

The ecosystems revolution is about you, me, and all of humanity, recognising that we are all co-dependent with one another, with our technology choices and their deployment, and with all of life with which we are integrally and unbreakably interconnected. The ecosystems revolution is one that progressively recognises that the wellbeing of the biospheric whole is the common 'mother lode' of wellbeing for all of humanity.

This necessary ecosystems revolution is what this book is all about and which, in the following pages, we will characterise and express in the language of opportunity.

Notes

1. Lovelock, J. (2000). *Gaia: A New Look at Life on Earth*. Oxford Paperbacks.
2. Ingham, E.R. (undated). *The living soil: bacteria*. United States Department of Agriculture. [online] http://www.nrcs.usda.gov/wps/portal/nrcs/detailfull/soils/health/biology/?cid=nrcs142p2_053862 (accessed 06 November 2014).
3. EarthSky. (2014). *How much do oceans add to world's oxygen?* (http://earthsky.org/earth/how-much-do-oceans-add-to-worlds-oxygen, accessed 16 June 2014).
4. MedicalXpress. (2014). Mouth bacteria can change its diet, supercomputers reveal. *MedicalXpress*, 12 August 2014. (http://medicalxpress.com/news/2014-08-mouth-bacteria-diet-supercomputers-reveal.html, accessed 06 November 2014).
5. Everard, M. (2011). *Common Ground: The Sharing of Land and Landscapes for Sustainability*. Zed Books, London. 214 pp.
6. Everard, M. (2013). *The Hydropolitics of Dams: Engineering or Ecosystems?* Zed Books, London.
7. Everard, M. (2015). *Breathing Space: The Natural and Unnatural History of Air*. Zed Books, London.
8. Cellulose, the primary constituent, has the chemical formula $(C_6H_{10}O_5)_n$, with a monomer molecular weight of 162. All of the six Carbon atoms (6× atomic weight of 12) and most of the Oxygen atoms (5× atomic weight of 16) derive from carbon dioxide (CO_2) gas.
9. 'No man is an island, Entire of itself, Every man is a piece of the continent, A part of the main. If a clod be washed away by the sea, Europe is the less. As well as if a promontory were. As well as if a manor of thy friend's Or of thine own were: Any man's death diminishes me, Because I am involved in mankind, And therefore never send to know for whom the bell tolls; It tolls for thee.' (John Donne 1572–1631)

10. Ceballos, G., Ehrlich, P.R., Barnosky, A.D., García, A., Pringle, R.M. and Palmer, T.M. (2015). Accelerated modern human–induced species losses: Entering the sixth mass extinction. *Science Advances*, 1(5), e1400253. DOI: 10.1126/sciadv.1400253.
11. Diamond, J. (2005). *Collapse: How Societies Choose to Fail or Succeed.* Viking Penguin.
12. Crutzen, P.J. and Stoermer, E.F. (2000). The 'Anthropocene'. *Global Change Newsletter*, 41, pp. 17–18.

CHAPTER 3

Breakthroughs in the Ascent of Humanity

Abstract 'Breakthroughs in the ascent of humanity' plots the trajectory of human development through the lens of the materials and technologies we have harnessed to further our own prospects. These have been characterised as a series of so-called 'revolutions' in the manipulation of natural resources. A generally narrow focus on immediate advantages accruing from largely fortuitous 'evolutionary' innovations has frequently also generated multiple unintended consequences, emphasising the need for greater cognisance of systemic ramifications for people and supporting ecosystems as a basis for the next societal revolution.

Keywords Natural selection • Artificial selection • Cultural evolution • Memes • Revolution • Innovation

The concept of and the term 'natural selection', famously introduced by Charles Darwin in his seminal 1859 book *On the Origin of Species*,[1] remains a cornerstone of modern biology. Natural selection describes how divergent traits become systematically either more or less common in a population due to the advantages or disadvantages they confer in any given situation. The power of the idea of natural selection lies in its recognition that heritable traits confer a higher probability of reproductive success on organisms adapted to interact more effectively with their environment, or alternatively, which lead to the progressive elimination of the less fit. The

© The Editor(s) (if applicable) and The Author(s) 2016
M. Everard, *The Ecosystems Revolution*,
DOI 10.1007/978-3-319-31658-1_3

fact that this was conceived at a time when there was no contemporary understanding of genetics, and so no valid theory of heredity, makes it an all the more remarkable conceptual leap.

The Human Evolutionary Journey

The ascent of humanity is fundamentally no different from the evolutionary process of natural selection. Our bodies may have changed little over the past 10,000 years. However, the recent phase of our evolution has seen massive advances in collective human understanding of the world around us and the wider universe, our capacities to communicate and innovate tools of increasing power, to express ourselves through increasingly diverse media, and to create cultures, governance, and belief systems. It is in the non-physical worlds that our evolution has blossomed to such as extent that we, at least in much of the developed world, extend our protection to those with physical disabilities because we see beyond their basic ability to compete biophysically, instead valuing their contribution to cultural diversity, knowledge, and creativity. Increasingly, we embrace, or at least seek to overcome prejudice against, different sexualities, races, and national affiliations as we strive to value the cultural contributions people may make rather than solely their competitive survival skills and hence perpetuation of genes.

For millennia then, the human journey has changed track from brawn to brain, elevating ideas, values, and capabilities as our evolving 'leading edge' and defining features. These attributes, of course, ramify directly back into physical reality—how we use, share, and dispose of natural resources—and so how we influence the natural world that so directly shapes our potential. Rather than acting as a purely physical beast at the whim of nature's vagaries, humanity has become innovator, maker, and user of ever more sophisticated tools, and communicator across time and space of knowledge, ideas, and artistic inspiration as our big forebrains conferred upon us significant differences, if not total distinction, from the rest of nature. Darwin also recognised that selection can be directed, drawing upon his knowledge of selective breeding of plants and animals put to human uses—artificial rather than natural selection—as a conceptual basis for analysing and understanding how natural forces can constitute natural selection. We have thus been exerting artificial selection pressures according to at least some of our preferences to shape attributes perceived as valuable to society.

The evolution of ideas is no less subject to natural selection than is modification of physical structure, wherein those best fitting a purpose tend to succeed preferentially to those less fit. The word 'meme' was coined by Richard Dawkins in his groundbreaking 1976 book *The Selfish Gene*.[2] 'Meme' is modelled on the word 'gene' but derived from a shortening of 'mimeme' ('imitated thing' in Ancient Greek from the same stem as the word 'mime'), introduced as a concept to address how evolutionary principles can explain the spread of ideas and cultural phenomena. Memes can constitute great ideas and technologies and novel uses of natural resources, but may as readily be applied to fashion, catchy tunes, or cultural habits and accepted norms. These ideas, behaviours, or styles spread from person to person, today increasingly through media, subsequently morphing, evolving, or dying within a culture. Memes thereby act as conceptual units that can shape imitable phenomena. Although the concept of the meme is far from universally accepted, its proponents regard memes as cultural analogues to biological genes in that they self-replicate, mutate, and respond to selective pressures.[3] As such, they are just as subject to the principles of natural selection, or at least artificial selection, going through processes of variation, mutation, competition, and inheritance, all of which influence their 'reproductive success'. Memes can prosper and differentiate, or else struggle and become extinct, just like biological traits and indeed whole species.

The human evolutionary tale then is in part defined by how natural forces drive us to react, but in increasingly larger measure in how we have learned to exploit those processes and pressures to our own advantage. And certain of those reactions have, in turn, led to dramatic breakthroughs in our capabilities.

HUMANS AND FIRE

The natural phenomenon of fire comprises a complex process of rapid oxidation of material, releasing heat and light, as well as a range of chemical substances. Fire has both positive uses and negative consequences for people, as well as significant implications for ecosystems across the world. Indeed, many terrestrial ecosystems are adapted to, or are dependent upon, regimes of fire which prevent them reaching a climax state thereby maintaining patches in different stages of succession.

A wide range of species of plants, animals, and microbes are adapted to exploiting these successional stages, fire contributing to overall biodiversity and functioning within the landscape. As just one of many examples, Canada's boreal forests comprise a mosaic of species and stands of deciduous, mixed deciduous-coniferous, and coniferous trees, the mix and distribution of which is controlled, along with their associated flora and fauna, by different intensities and durations of fire over long periods of time.[4] Species respond differently to fire, for example, with forest regeneration on burned sites beginning with the establishment of pioneer species that are well adapted to landscapes where fires recur regularly. Some tree species in the boreal forest re-establish quickly by sprouting from charred stumps and roots. Other tree species re-colonise quickly by producing abundant seeds, with some species, such as jack pine and lodgepole pine, requiring the heat of fire to release their seeds. Fire also releases nutrients from the soil, eliminates competing species, and opens the canopy allowing in sunlight to the forest floor, promoting the germination and fast growth of saplings. By contrast, balsam fir and cedar are less well adapted to withstanding extensive fires, so tend to be rare in areas that are repeatedly severely burned or where fires are large. Consequently, fire serves as a vital ecological component of these extensive Canadian forests, as indeed many other ecosystems, with their effective conservation therefore depending on maintaining or managing natural fire regimes. The principle of fire as a controlling agent maintaining diversity applies equally to savannah and many grasslands, other forests, and a range of different habitats across the world, within which various co-adapted plant and animal species make use of fire and its aftermath for competitive benefit.

However, as evocatively put in Rudyard Kipling's *The Jungle Book*,[5] the harnessing of man's 'red flower' marks a key distinction in cognitive and manipulative ability between humanity and the myriad species with which we co-exist. The point, or points, at which humanity learned to control fire is generally considered to have occurred during the Neolithic period, and had many implications ranging from the generation of heat and light to the manipulation and appropriation of the productive capacities of whole ecosystems. Just one example of the benefits to humanity of harnessing the power of fire was that it enabled people to cook their food as long ago as 1.9 million years ago, increasing the variety of food sources and the availability of nutrients. Fire also enabled people to stay warm in cold weather, to live in cooler climates, and to scare away nocturnal predators. Other diverse uses of fire throughout history include in landscape management, for example, to control scrub and to force new grass growth. Burning techniques are still widely used today for such purposes as forcing the growth of new grass for grazing, for 'cool burns' to encourage the growth of timber crops ridded of low-growing competitive vegetation, and in 'slash-and-burn' agriculture which remains common throughout tropical Africa, Asia, and South America. Fire is also used as a weapon, for torture and execution, and for the controlled harnessing of energy in electricity generation, steam power, and internal combustion engines.

However, it is not merely the harnessing of fire that has provided humanity with so many advantages. Other human 'revolutions' have been defined in terms of step changes in our capacities to exploit physical materials.

MATERIAL REVOLUTIONS

The Stone Age describes a broad period of prehistory spanning approximately 3.4 million years, during the period between 6000 and 2000 BCE (an abbreviation for Before the Common, Current or Christian Era according to different definitions), during which, stone was widely used by people to make tools with sharp edges, points or percussive surfaces. Artefacts from the Stone Age include a variety of tools formed from a diversity of types of stone ranging from flint and chert for cutting and weapons, basalt and sandstone for grinding, natural substances including bone, shell and antlers and, in the later period, fired clays for pottery. Although other materials were used, stone tools were not only commonplace but also survived better in archaeological records.

The Stone Age coincides with evolution of the genus *Homo*, although some genera such as *Australopithecus* and *Paranthropus* preceding and contemporaneous with *Homo* may also have manufactured stone tools. *Homo erectus*, the predecessor of modern humans, found an ecological niche in the savannah of the Rift Valley, and was defined significantly by its capacity to make and develop tools. The Stone Age is defined primarily by the durable artefacts it left behind, crudely and commonly divided into three phases: the Early Stone Age, Middle Stone Age, and Later Stone Age. However, the Stone Age saw a progressive evolution in tool sophistication and human capabilities, including development of agriculture and the domestication of some animals, with parallel advances in social structure and traditions, range of food sources and capacity to exploit new environments. Diverse other aspects of cultural practice and evolution can be assumed, and indeed some are evidenced by the kinds of tools these people left behind.

The Stone Age period ended not, as noted before, because people ran out of stones, but with the advent of metalworking enabling people to develop metal tools better fitting their needs and extending their capabilities. Transition to the age of metal was incremental, stemming from the innovation of techniques for smelting ore. It was also far from even, the discovery and practice of metal smelting varying significantly across the ever-expanding geographical range of humanity.

The Bronze Age, another long evolutionary journey for humanity spanning approximately 3300 to 1200 BCE, denotes an era defined by the first significant metal manufactured this way, bronze formed as an alloy of copper and tin. There was a transitional period, known as the Copper Age, during which our ancestors had learned to smelt copper but not yet to synthesise bronze. Some scientists classify the Bronze Age as the second principal period of the three-age Stone-Bronze-Iron system, some civilisations smelting their own copper and alloying with tin whilst others traded bronze for other products, perhaps due to the rarity and uneven distribution of copper-tin ores in surface layers of the Earth's crust. The Bronze Age witnessed significant social evolution, including the rise of Mesopotamian and Egyptian cultures each of which developed the earliest viable writing systems, as well as inventions such as the potter's wheel, centralised government, codes of law, the building of empires, societal differentiation, slavery, warfare, science, and mathematics. This explosion of cultural evolution spread from North Africa to Central Asia, East, South, and Southeast Asia, Europe and some civilisations in South America. The Bronze Age was late to arrive in Japan and sub-Saharan Africa.

There was a progressive transition from the Bronze Age to the next age—the Iron Age—defined by the prevalent use of iron and some steel tools used for cutting and weapons. The earliest known iron artefacts have been dated to 3200 BCE in Gerzeh, northern Egypt, made from meteoritic iron shaped by careful hammering.[6] Ancient inhabitants of parts of Niger are thought to have become the first iron-smelting people in West Africa and amongst the first in the world at around 1500 BC. However, the spread of understanding of iron metallurgy, including purification of iron from oxidised iron ores and the consequent wider use of iron objects proliferated rapidly and widely across the human world between 1200 BCE and 1000 BCE. Iron is barely harder than bronze but, when combined with carbon, the resultant steel is far harder. Shortage of tin may also have contributed to the transition to more abundant and accessible iron, and the stronger and lighter products made from it.

This technical sophistication coincided with a range of wider changes in society including progressive agricultural practices, religious beliefs, and artistic styles. Principal features distinguishing the Iron Age from preceding ages include introduction of alphabetic characters as a written language, literature, and historic records; the Iron Age saw some of the earliest Sanskrit, Chinese, Indian Vedic, and Hebrew Bible texts preserved in the manuscript tradition. The commencement of the Iron Age in Europe and adjacent areas enabled a proliferation of tools, weapons, ornaments, pottery, and decorative designs. Again, the dates and context of proliferation of the Iron Age varied by region across the constantly expanding human world.

The genesis of and transitions in the Stone-Bronze-Iron system—describing the materials used by society throughout a long journey of societal evolution spanning approximately six millennia—were not directed, but resulted from iterative progressions stemming from discovery and manipulation of materials better suited to evolving human activities. Each phase was a progressive journey of innovation and sophistication in material use, a form of artificial selection based on greater fitness for purpose, but also a trigger for wider innovations, capabilities, and cultural complexity.

TECHNOLOGICAL REVOLUTIONS

The Stone-Bronze-Iron journey is, of course, intersected by other technological revolutions of various kinds, in which manipulation of a variety of other materials and biological resources was also pivotal.

As outlined in considerable detail in my 2011 and 2013 books *Common Ground* and *The Hydropolitics of Dams*,[8] innovations in the manipulation of water and its implications for the productivity of land had a profound role in ushering in successive waves of cultural evolution. This was certainly true of the first recorded civilisation in Uruk, in the 'Fertile Crescent' of Mesopotamia during the Bronze Age around 5300 BCE. The history of enhanced food productivity, and its contribution to the settlement and differentiation of successive civilisations liberated from the drudgery of foraging for food, relates significantly to controls of water and the flows of nutrients and other substances that it conveys.

Advantages stemming from the rise of 'hydraulic civilisations', defined by the historian Karl Wittfogel as a social or government structure maintaining power and control through exclusive control over access to water,[9] have underpinned many subsequent civilisations. Innovations in control of water then promoted profound revolutions underpinning those more normally attributed to agriculture, with control of the flows of water subsequently constituting an often underappreciated but frequently integral part of the rise of nations, empires, and civilisations, as a medium for production but also sometimes for oppression and warfare. Some productive water management technologies, such as traditional Asian paddy and terracing systems, have endured for millennia as an efficient means to exploit local water resources, with low environmental impact. The clear advantages and sustainability of this technology has resulted in its spreading across much of the tropics, enduring and forming a uniting theme for local people whilst civilisations have risen, fallen and been replaced around them. And so the arc of water management has continued to rise in sophistication, quantity, and geographical range, now including the re-plumbing of entire continents such as massive water diversion and inter-basin transfer schemes particularly in China.

Control of water, and of the nutrients and other constituents that it bears, ushered in new waves of cultural evolution as people shared learning about harnessing natural flows and processes for their own ends, freeing themselves from the drudgery of daily hunting and foraging for food and water and enabling settlement and subsequent differentiation of cultures. This era of human history is sometimes referred to as the 'Agricultural Revolution', which witnessed widespread transition of many human cultures from hunter-gatherer lifestyles towards agriculture and settlement from around 12,000 years ago (and so clearly contemporaneous with the Stone-Bronze-Iron narrative). This profound and wide-scale revolution in agriculture comprised adoption of novel food-producing techniques, radical modification of the natural environment through techniques such as irrigation and deforestation, domestication of some animal and plant species, and emerging technologies such as food storage. These innovations progressively formed a basis for sedentary lifestyles, the founding of villages and towns served by manipulation of, rather than foraging for, natural resources and, eventually, the burgeoning of cities within which monuments, writing systems and arts prospered enabling significant differentiation of societal roles.

The initial Neolithic 'Agricultural Revolution', or Agrarian Revolution, was in reality the first in many waves of such 'agricultural revolutions' throughout human history. Others have included the Arab Agricultural Revolution, occurring between the eighth and thirteenth centuries AD, during which innovations in crops and farming techniques spread across the Arab and Muslim worlds during the Islamic Golden Age.

More familiar to those educated in the western world will be the British Agricultural Revolution, generally described as approximately between 1750 and the end of the nineteenth century, although in reality tracing far earlier roots. These include, for example, innovation and subsequent spread from around 1590 of the water meadow system, entailing sophisticated conversion and management of floodplains, particularly in the river catchments of southern England, as a means to capture the flows of moisture, nutrients and, above all, warmth from rivers.[10] This enabled estate owners and managers to force the growth of new grass during the 'hungry gap'—the period between depletion of the preceding year's hay reserves and the seasonal re-emergence of fresh spring grazing—which hitherto imposed a limitation on livestock for food, fibre, traction, transport, and wider economic activities. There were further linked innovations and added values, such as the sheep-corn system under which sheep grazed on water meadows by day and were driven to downland tops by night, where their bodily wastes fertilised the sparse soils, significantly increasing the productivity of wheat and other arable products. So significant were their advantages that water meadows proliferated within a few decades to almost everywhere across England where catchment topography and geology were favourable. The British Agricultural Revolution saw a substantial increase in agricultural productivity in Great Britain, in turn helping drive the subsequent Industrial Revolution (to which we will turn shortly).

The Scottish Agricultural Revolution refers to a period between 1760 and 1830 during which the British Agricultural Revolution spread north into Scotland, particularly leading to the Lowland Clearances, which commercialised and substantially changed the traditional system of agriculture in Lowland Scotland.[11] One consequence of this was inflation of rents, pricing many tenants out of the market and replacing part-time labourers or sub-tenants (known as cottars, cottagers, or bondsmen) with full-time agricultural labourers thereby profoundly changing the way of life in many parts of Southern Scotland. Migrating from traditional homelands that could no longer sustain their livelihoods, thousands of these displaced cottars and tenant farmers migrated to emerging industrial centres such as Edinburgh, Glasgow and other burgeoning cities across the northern UK and further afield overseas for employment in the early Industrial Revolution.

Agricultural innovation has not ceased, nor stopped spreading across the world. Another period of substantial change in agriculture, known as the Green Revolution, occurred in the latter part and following the Second World War. Between 1943 and the late 1960s, concerns about food security drove substantial investment in a sequence of research, development, technology transfer, and commercialisation activities that significantly increased industrialised agricultural production worldwide. Unfolding of the Green Revolution, also sometimes referred to as the 'Second Agricultural Revolution', began most markedly in the 1960s, with a range of initiatives such as development of high-yielding cereal crops, expansion of irrigation, modernisation of management techniques, distribution of hybridised seeds, and increased innovation and use of synthetic fertilisers and pesticides. This latter 'Green Revolution', a term first used in 1968 by William Gaud, former Director of the United States Agency for International Development, contrasted the spread of these new technologies in characteristically Cold War terms with the violent Soviet Red Revolution and the Shah of Iran's White Revolution.[12] The 'Green Revolution' is credited with saving over a billion people from starvation.[13] However, for all the benefits that the Green Revolution brought to humanity in terms of food sufficiency, there is broad consensus that it did much massively to reduce agricultural biodiversity, reliant as it was on just a few high yield varieties of each crop and stock, with equally severe implications for the depletion of wild biodiversity.[14] The consequences of this erosion of biodiversity include not only increased food supply vulnerability due to increased risks of epidemics sweeping through a depleted gene pool, but also potentially a serious reduction in the functionality and resilience of ecosystems and the flow of multiple ecosystem services underpinning human wellbeing and opportunity.

Perhaps the best-known 'revolution' in the western world is the European Industrial Revolution, a term applied to describe a long-term transition to new manufacturing processes from about 1760 to sometime around 1840. This was indeed an era of remarkable innovation, including for example transition from production systems based on manual methods to increasing reliance on machines, and from animal power and biological fuels towards alternative sources including water and more energy-dense fuels such as coal. There were also innovations in a range of chemical processes significantly including advances in iron-making and other aspects of metallurgy, as well as development of novel machine tools, inventions such as cement and gas lighting, new methods of glass-making and paper-making, and advances in transport systems. These changes elevated the volume and profitability of the dominant textile industry, the first to use modern production methods with such innovations as the 'spinning jenny' and the 'water mule', but eventually reached into many other sectors of human enterprise. Much as the first recorded settlement in Uruk, this new explosion in human technical capability revolutionised almost every aspect of daily life from average per capita income, instigation of capitalism and consumerism, and dissemination of knowledge through mechanised printing. The Industrial Revolution also conferred unprecedented wealth on a minority, many of whom were not necessarily favoured by birth into privileged classes—though the benefits of industrialisation were far from evenly distributed across society—spurring grand philanthropic gestures such as investments in public health and education, philosophical and scientific enquiry and discovery, and exploration and the building of empires in the quest for more resources to satisfy the demands of burgeoning industry.

Other technological revolutions, even if not so regarded, have been no less miraculous. Advances in medicine have been and continue to be dramatic, from the discovery of germ theory and innovations to fight, and in some cases even to eradicate, some forms of communicable diseases. Advances in drugs, both natural substances and their synthetic analogues such as aspirin—the first nonsteroidal anti-inflammatory drug—have created massive breakthroughs in human health and comfort. Discoveries of antibiotics, of medical imaging technologies such as X-rays, Magnetic Resonance Imaging, and Computerised Tomography, and many more breakthroughs besides have been nothing short of the stuff of relatively recent science fiction.

We have lived through the Space Age, with its massive breakthroughs in propulsive, remote sensing, and broader aeronautical capabilities. The innovation of integrated circuits (ICs)—those little, ubiquitous silicon chips that we take so much for granted as the 'brains' in our watches, televisions, central heating timers, computers, tablets, and phones, and seemingly increasingly everywhere else—date back to 1949, arriving in the form familiar today through successive waves of innovation each conferring selective advantages. ICs comprise a complex set of electronic circuits etched into a small plate, or 'chip', of silicon or other semiconductor material. A chip no bigger than a fingernail may today contain several billion transistors and other components, connected by electronic tracks that may be only tens of nanometres wide. The discovery that assemblages of semiconductors could perform functions formerly performed by vacuum tubes, enabling not merely miniaturisation but also very substantial cost reductions and massively more robust, replicable, and complex compound circuits, led to the advent of powerful miniature computer functions. Early ICs were crucial to the progress of aerospace projects, to the extent these projects were some of the most significant drivers of further development of IC technology. Indeed, the need for lightweight digital computers for guidance systems for the Apollo space missions served as a powerful motivation for major step changes in the evolution of IC technology.

ICs are now used in virtually all electronic equipment and have revolutionised many other fields of technology. Indeed, innovation of ICs has made possible much of the revolution in Information Technology, which has in turn changed the world in formerly unforeseen ways. The IT revolution has put instantaneous mobile communications, powerful computation and access to the sum total of human knowledge into the hands of virtually everyone. This advanced IT capacity has also revolutionised banking, publication, terrorism and defence, sharing and analysis of medical records, creative arts and the distribution of music, resource efficiencies, shopping, weather forecasting, and day-to-day telecommunications. We live in wondrous times, in which Moore's Law—the observation that the number of transistors in dense ICs doubles approximately every 18–24 months so triggering proportionate increases in the capabilities of many digital electronic devices simultaneously with decreasing microprocessor prices—still continues to hold true, highlighting that this disruptive journey is very far from exhausted.

Many more types of 'revolution' have contributed to the progressive journey of humanity, with examples across many walks of life, albeit that their benefits remain unevenly distributed across global society. Chemical innovations are some of the more pervasive, ranging from development of fertilisers, explosives, propellants, lubricants, drugs, scents, and many more beneficial, though also potentially hazardous, applications. Of this long history of chemical invention, development of plastics deserves special mention as revolutionary in their own right, as well as making possible revolutions in other fields. The term 'plastic' describes any of a wide range of synthetic or semi-synthetic materials that are malleable, potentially moulded into solid objects that can vary in rigidity and shape, generally based on long-chain organic polymers containing a range of embedded additives that further modify their properties. The term 'plastic' was coined by Leo Baekeland, inventor in 1907 of the world's first fully synthetic plastic known as Bakelite. Over the following century and more, a hugely diverse range of plastics has been developed that confer a range of benefits to society due to properties such as their durability, electrical and thermal insulation, relatively low cost, ease of manufacture, versatility, imperviousness to water and weathering, and capacity for their properties to be further modified through inclusion of additive chemicals such as plasticisers, stabilisers, pigments, and impact modifiers. This has led to the pervasion of plastics across a range of applications—from document wallets and toys through to furniture and fabrics, paints and packaging, and medical and electronic equipment—progressively displacing traditional materials such as metal, leather, wood, stone, glass, and ceramics. The packaging, construction, and automotive sectors are particularly significant users of plastics by volume in developed countries.

Plastics have also made possible breakthroughs in a range of other fields, such as their role in radar technology acknowledged as significantly influencing the pace and outcomes in the Second World War, as well as in electronics and small-scale engineering where their consequences for significant advancements across many other spheres of human interest, from banking to communications and aerospace, have been at least as creatively disruptive as ICs. Plastics have also enabled major advances in medicine, with applications ranging from polymer implants to readily sterilised surfaces and medical imaging equipment. Like most revolutions, comprising evolutionary steps based on immediate advantage, the wider ramifications of plastic manufacture, use and disposal have been substantially overlooked in the journey since the early twentieth century towards their current pervasion. A range of environmental and health concerns has arisen over time as awareness has emerged of issues associated with chemical pollutants potentially released, particularly during manufacture and disposal as well as the accumulation of plastic litter, particularly in urban areas and in oceanic gyres, due to their slow decomposition. Towards the end of the century, this growing concern has driven innovation around recycling and recyclability as a wider quest for their sustainable use, a further, more wilfully directed stage in the revolutionary journey as we shall see when reviewing progress across the European polyvinyl chloride (PVC) industry in Chap. 4

THE ASCENT OF HUMANITY

Humanity has ascended steeply towards our current profusion, global pervasion, and emancipation from the restrictions of predation, starvation, disease, desiccation, and environmental extremes. This has been empowered by cascades of discoveries offering competitive advantage, evolving through successive innovative additions to an inherited body of knowledge, progressively to enable us to harness the energy of water, air flows, and chemical and nuclear reactions, to exploit materials as diverse as rock, clay and aggregates, metals, nutrients, fossil carbon reserves, radio-nuclides, and semiconducting materials, and to appropriate a substantial proportion of nature's productive capacities for the sole advantage of our species.

Our technological reach has extended to include not just harnessing nature in situ but, as graphically demonstrated by the ascent of hydraulic civilisations, diverting whole massive flows of water to where we choose to live, farm, mine, and manufacture. We trade and transport food and other commodities across catchments and continents, and we also pack-age and convey energy in the form of transportable oil and other fossil fuel deposits, in batteries and fuel cells, and in the form of hydrogen and other energy carriers. We even 'package' natural metabolic and productive processes for our utility in the form of agricultural, horticultural, aquacul-tural, and sewage treatment technologies.

Much historic literature, and a great deal of contemporary common perception, may frame this as evidence of humanity's superiority over and apartness from nature. Some people even wrap this sentiment in religious assertions that generally serve often only to divide different chosen beliefs. However, all of these impressive breakthroughs, in fact, provide evidence of our ultimate dependence on nature. All this conscious cognitive capac-ity and co-constructive synthesis and application of knowledge, unique in the biosphere's evolutionary timeline to date, always relating in some way to more empowered exploitation of natural assets and processes. Each breakthrough, in fact, confirms our dependence upon nature for all facets of our health, wealth creation, and wider wellbeing, merely representing technical means to harness nature's beneficence in more accessible, con-centrated, convenient, and lucrative forms.

Our pathway of development to date, for all its innovative genius, has generally proceeded with a narrow focus on immediate rewards yet with a far from competent grasp of the principles of co-dependence with the planetary ecosystems that underwrite it. It has consequently tended to overlook a plethora of negative implications arising from contemporary lifestyles for nature's supportive and productive systems, which form the irreplaceable underpinnings of a secure future. As nature's supportive eco-systems have become undermined by many innovative yet short-sighted breakthroughs shaping the ascent of humanity, threats to the security and continuity of human wellbeing, and indeed the integrity of the whole biosphere of which we are an integral part and share a common destiny, have increasingly manifested as a now familiar suite of sustainability pressures.

Our trajectory of technological progress has delivered, for those privileged by birth to be part of the already developed world, serially improving prospects for material quality of life. However, the metabolism of that model of progress is beginning to limit future prospects through its wide range of unintended consequences for natural resources and wider environmental security. Today, the 'millennial generation' is likely to be the first unlikely to enjoy improving life prospects and quality, a potent warning of an urgent need for a new vision and model of future human development spanning developed and developing worlds alike. Anthropogenic climate change is one of the most profound forces likely to impose further self-limitation, potentially unstoppably, as a form of 'nature's revenge' that is already evidently under way.[15]

Assurance of humanity's continuing ascent demands breakthroughs of a rather different character. What we need now are step changes selected on the basis of growing synergy with the biospheric supportive processes that constitute the ultimate foundational resources underwriting a progressive future. This next revolution will be framed by increasing symbiosis between the breadth of societal activities and the ecosystems vital for continued human security and opportunity.

NOTES

1. Darwin, Charles. (1859). *On the Origin of Species by Means of Natural Selection, or the Preservation of Favoured Races in the Struggle for Life.* London: John Murray.
2. Dawkins, Richard. (1976). *The Selfish Gene.* Oxford University Press.
3. Symons, Donald. (1979). *The Evolution of Human Sexuality.* Oxford University Press.
4. Natural Resources Canada. (undated). *Fire ecology.* (http://www.nrcan.gc.ca/forests/fire/13149, accessed 11 November 2014).
5. Rudyard Kipling. (1894). *The Jungle Book.* Macmillan Publishers, London.
6. Rehren T, et al. (2013). 5000 years old Egyptian iron beads made from hammered meteoritic iron. *Journal of Archaeological Science*, 40, pp. 4785–4792.

7. Everard, M. (2011). *Common Ground: The Sharing of Land and Landscapes for Sustainability.* Zed Books, London. 214 pp.
8. Everard, M. (2013). *The Hydropolitics of Dams: Engineering or Ecosystems?* Zed Books, London.
9. Wittfogel, K. (1957). *Oriental despotism: a comparative study of total power.* New York: Random House.
10. Everard, M. (2005). *Water meadows: living treasures in the English landscape.* Forrest Text, Ceredigion.
11. Aitchison, P. and Cassell, A. (2003). *The Lowland Clearances: Scotland's Silent Revolution 1760–1830.* Tuckwell Press Ltd, Edinburgh.
12. Gaud, W.S. (1968). *The Green Revolution: Accomplishments and Apprehensions.* AgBioWorld, 8 March 1968.
13. Hazell, P.B.R. (2009). *The Asian Green Revolution.* International Food Policy Research Institute (IFPRI) Discussion Paper. (https://www.ifpri.org/publication/asian-green-revolution, accessed 12 October 2015).
14. Kilusang Magbubukid ng Pilipinas et al. (2007). Victoria M. Lopez, et al., ed. *The Great Rice Robbery: A Handbook on the Impact of IRRI in Asia.* Penang, Malaysia: Pesticide Action Network Asia and the Pacific in collaboration with Sibol ng Agham at Teknolohiya, Inc. (http://www.panap.net/en/r/post/rice/193, 22 November 2014).
15. Pearce, F. (2006). *The Last Generation: How Nature Will Take Her Revenge for Climate Change.* Eden Project Books, Cornwall.

CHAPTER 4

Chance or Choice?

Abstract 'Chance or choice?' reviews the nature of natural selection, and how selection principles also apply to the evolution of ideas, technologies, and products. The multifactorial nature of natural selection is contrasted with the generally narrow framing of artificial selection, which externalises many impacts on ecosystems thereby compromising their capacities to sustain human needs through the process of natural selection. A more directed revolution is required in human development, rather than reliance on fortuitous innovations, guided by a broader framework of principles more closely aligned to the complexity of the natural world. Conceptions of sustainable development implicitly assuming stationarity are challenged, as the pace of ecosystem declines and burgeoning of human numbers demands a concerted approach to rebuilding degraded ecosystem capacity and resilience.

Keywords Natural selection • Artificial selection • Choice • Foresight • Directed and undirected change • Backcasting

A central building block of Charles Darwin's theory of evolution was comparison of artificial selection of favourable traits in domesticated species, particularly varieties of pigeon, with selection by natural processes. Artificial and natural selection differed, in Darwin's analogy, only to the

© The Editor(s) (if applicable) and The Author(s) 2016   35
M. Everard, *The Ecosystems Revolution*,
DOI 10.1007/978-3-319-31658-1_4

extent that it was the agency of humans rather than of nature that 'chose' favoured features arising through natural variability.

NATURAL SELECTION

Although the term 'Survival of the fittest' is commonly invoked to describe the process of natural selection, it is not coined one or, it seems, initially favoured by Charles Darwin. Indeed, the phrase did not appear in *The Origin of Species* until its fifth edition. Rather, the sense that Darwin initially communicated was that survival to pass on heritable traits—and this, let us recall, was before any viable theory of how inheritance of traits might occur—was favoured in individuals best adapted to their environment.

Publication of *The Origin of Species* was famously delayed due to Darwin's concerns about how the notion of progressive evolution of species by impartial biological pressures conflicted with the dominant theocentric, creationist world view. Evolutionary biologist Richard Dawkins had no such reservations in taking a gene-centric view of the world, a perspective not available to Darwin, in his 1986 book *The Blind Watchmaker*,[1] elaborating further on how complexity arises without the intervention of a 'creator'. By the 'trial and error' of survival or extinction, or at least the favouring of more or less well-adapted variations, species morph or diverge progressively over time.

All organisms obey simple biological principles. One of these is observed in how populations respond to resource limitation. Classical biological theory, corroborated by the growth of microbes in laboratory cultures but applicable to all biological systems, is illustrated by Fig. 4.1. In this model, organisms tend initially to grow logarithmically until a point of resource limitation (curve 1a), after which death ensues and the population declines (curve 1b). However, if new resources are introduced into the system, and assuming no other resource is now limiting or that toxic waste products do not impose their own limitation, a renewed cycle of growth ensues (curve 2a) up to the point of depletion of the new resources, beyond which population once again goes into decline (curve 2b). Alternatively, more resources may be added, resulting in subsequent growth to the point of limitation (curve 3a), and so on.

The evolution of species and ecosystems has progressed by selection of favourable adaptations to exploit new niches or resources, or to overcome those imposed by predation, parasitism, shifts in climatic regime, and a host of other environmental pressures.

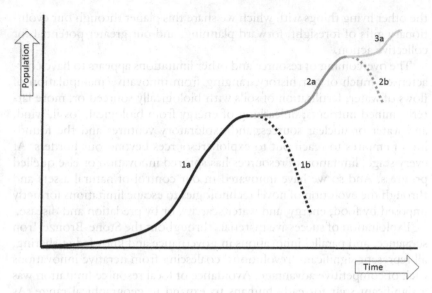

Fig. 4.1 Classical growth curves of organisms exploiting and depleting resources

SELECTION IN HUMAN ENDEAVOUR

Humans are no different from other organisms in terms of our dependence on natural resources as diverse as air, water, food and its complex of embedded nutrients and energy, and potential toxicity from accumulation of waste products. Much of the progress of humanity can be explained by the many ways by which our species has taken control of its future—biophysically and culturally—through innovations of selective advantage in overcoming resource limitations. These exhibit the same principles of selection of favourable adaptations across many fields of human endeavour. Resource depletion or exclusion from supply, for example on cost or political grounds, or the 'toxicity' of adverse corporate reputation amongst consumers all conspire in markets to limit growth leading to an inevitable collapse and 'death'.

Where humanity differs—cognitively and commercially—from other organisms is in our radically greater capacity to exploit alternative resources, the 'never say die' survival instinct we share with all living things manifesting culturally in innovations to overcome resource limitations and other natural threats. This is often achieved in unique ways compared to the other living things with which we share this planet through our evolutionary gifts of foresight, forward planning, and our greater potential for collective action.

The overcoming of resource and other limitations appears to have characterised much of our history, ranging from innovative manipulation of flows of water; fertilisation of soils with biologically sourced or, more latterly, mined nutrients; unleashing of energy from biological, fossil, wind, and water or nuclear sources; and exploratory ventures and the founding of empires to reach out to exploit resources beyond our borders. At every stage, limitation of resources has spurred innovation or else quelled progress. And so we have innovated in our control of natural assets and through the evolution of novel technologies to escape limitations formerly imposed by food, energy, and water scarcity, or by predation and disease.

Exploitation of successive materials throughout the Stone-Bronze-Iron sequence, and parallel innovations in governance and knowledge-sharing, all represent significant 'revolutions' coalescing from iterative innovations each of competitive advantage. Avoidance of local resource limitation was a significant spur for early humans to expand in geographical range. As the European Industrial Revolution exploded, hunger for more abundant and cheap resources beyond those of depleted homelands propelled many European nations into an era of empire-building. The same behaviours are observed today as China reaches out to take control of food production, increasingly scarce mineral and other resources in Africa, India, and other territories, also embarking on mass damming of transboundary rivers to appropriate their flows of energy and water preferentially over downstream neighbours. These adaptations have cumulatively helped humanity evade limitations of one form or another, just as historic and ongoing industrial, agricultural, IT, and other revolutions extend our capabilities and advantages.

We see the same selection pressures in business weeding out 'less fit' products and services, as innovations better serving a purpose, including those that are cheaper or otherwise more favourable, become preferentially selected by open markets. For example, the days of the phonograph—a groundbreaking device invented in 1877 by Thomas Edison to reproduce sound recorded onto a tinfoil sheet phonograph cylinder—are now long behind us. This technology, a breakthrough that must have appeared near-miraculous in its day, was displaced through the 1880s by wax-coated cardboard cylinders enabling both mechanical recording and reproduction of sound. Further 'competitive advantage' was realised subsequently by grooved disks in gramophones or later on record players that, in time, also became electrified. But the '78' and its successor LP, or long-playing record, remain with us now only as relics gathering dust in the attic, or else as the nostalgic preserve of vinyl aficionados. The days of the cassette tape and the minidisc too, each in their time offering breakthroughs in miniaturisation, portability, and robustness, have come and gone. The CD, or Compact Disc, too is waning in sales, as more music today is either downloaded in MP3 or other digital formats or else streamed in real time through the Internet. The desire to listen to music and the spoken word remains, but the business case driving the delivery medium has advanced through formerly unexpected technological leaps conferring the selective benefits of greater utility, accessibility, fidelity, and value to the listener.

We could make similar analogies with the transition of video entertainment from broadcast only, to video cassettes and laser disks each witnessing interesting 'survival of the fittest' battles between competing formats (such as VHS versus Betamax video cassette tapes), thence to digital DVD and on to live streaming of video content on computers, smart TVs, and now on mobile hand-held devices.

However, as the burgeoning human population runs into increasing conflict with dwindling natural resources, we have to recognise that, in a planetary system with finite boundaries, we cannot keep expecting to reach out for fresh resources to overcome limitations to recharge the metaphorical laboratory culture of human microbes. And we are finding out somewhat belatedly that, in overcoming one limitation or innovation, our pathway of technological innovation conferring specific biophysical or market advantages has been blind to collateral implications for other closely connected, historically disregarded natural limiting factors.

We are, for example, walking into a future of 'peak oil', in which the dwindling availability of economically exploitable fossil fuel reserves is being outstripped by burgeoning global demand, at the same time remobilising carbon sequestered from the atmosphere over geological timescales with alarming implications for climatic stability. We have also historically overlooked how our innovation of refrigerant and propellant chemicals has yielded major benefits for food storage and comfortable lifestyles, yet has led to the unintended consequence of catalytic breakdown of the vital stratospheric ozone shield. Equally, advances in electromechanical technologies to control sewage pollution to achieve the laudable and desirable goal of reducing pollutant loads entering watercourses have inadvertently elevated fossil fuel-based energy inputs with associated carbon-rich aerial emissions, use of increasingly scarce mined materials that often come from politically troubled regions, and also the need to dispose of slurry and other forms of solid waste. We are also stumbling into a future of 'peak phosphorus' in which global economically exploitable reserves of this vital addition to modern food production systems are under greater pressure from international trade, whilst the downstream effects of nutrient pollution are fundamentally changing terrestrial and aquatic biodiversity and system characteristics and with it the balance of benefits and threats flowing from them.

The problem with iterative, issue-by-issue evolutionary technological progress is that it is generally blind to systemic consequences, overlooking implications for non-focal social and environmental factors over the longer term. This contrasts starkly with natural selection, in which systemic connections to whole ecosystems impose a complex of simultaneous selection pressures. The blinkering of market-based selection to narrowly framed and often immediate benefits potentially creates many more problems than it solves in terms of what is omitted from commercialisation, regulation, and other decision-making and operational processes. The host of sustainability challenges facing humanity today highlights the failure of narrowly framed and short-term market selection to account for its multiple unintended consequences for the ecosystems ultimately underpinning continuing human wellbeing and progress, putting at risk continuing security and opportunity, as the impartial and systemic forces of natural selection come to bear. In truth, we run the real risk of reaching a point when systemically interconnected natural selection pressures bear down upon wider unaccounted ramifications stemming from a blinkered set of market-defined artificial selection pressures, imposing terminal limitations on our lifestyles and potential to live healthy and fulfilled lives.

We have observed how our evolutionary gifts of foresight, forward planning, and collective action have propelled us through historic limitations, successive innovations cascading into revolutions that have then formed the bedrock of whole civilisations. That same human spirit is needed now to reimagine and create a new era of evolution, reconnecting us to the finite yet inherently renewable supportive capacities of the world we inhabit and the consequences of impartial natural selection pressures on contemporary lifestyles. However, the character of future innovations and artificial selection has to broaden in scope, progressively taking a more systemic approach that accounts for and seeks greater symbiosis with the natural supportive processes upon which future wellbeing depends.

OUR CHOSEN FUTURE?

Various dictionaries define 'revolution' as either completion of a circuit or as achievement of a new set of norms. Sometimes, there is a glamorous emphasis upon change brought about by force, so let us at this point put aside this negative connotation. Armed uprisings popularised, and as often sanitised, in the political telling of history are no guarantors of stable and democratic change. Rather more complex, less romanticised realities are entailed in sustainable regime change, necessarily bloodless and consensual if they are to endure as the kind of fundamental, sector-spanning paradigm shifts that have characterised humanity's interrupted ascent to date.

Most enduring revolutionary regime changes in practice comprise cumulative evolutionary steps. In my book *Breathing Space*,[2] I observed that *'Revolutions are generally evolutionary processes, each innovation progressing on the basis of prior inventions over a period of time dramatically compressed by hindsight'*. For example, only with hindsight can the European Industrial Revolution be regarded as revolutionary, actually comprising cascades of iterative, undirected change occurring most intensively over a period of a century. Likewise, the European Agricultural Revolution spanned over three centuries of incremental innovation rather than orchestrated change. Industrial and agricultural innovations continue today, with some countries just embarking on their own 'revolutions'. Many other such revolutions, from the harnessing of energy, materials and biological productivity, as indeed in technological breakthroughs and market regime shifts, highlight how revolutions build from cascading evolutionary stages, each offering selective advantage.

There is also a distinction between revolutions that are undirected— at the whim of happenchance—and those forged more wilfully. In other words, some are chosen and directed, whilst many more have been achieved by progressive, selective, or otherwise undirected steps over long periods of time that, only in retrospect and when condensed by historical perspective, are perceived as combining into a coherent revolution.

Examples of undirected changes which nevertheless build into a great degree of complexity and regime shift, akin to the neo-Darwinian, gene-centred world view lacking the intervention of a 'creator', can be found in the arising of biological diversity and complexity but also in examples as mundane and recent as those described in the evolution of sound reproduction and video technology. Many undirected revolutions have resulted from cumulative changes that happen to better fit a human need or a changing market, ecosystem or other selection pressure. As we shall see in Chap. 6, undirected change is also reflected in fragmented progress over the past century in embedding selected environmental and ethical concerns into new societal norms that, although welcome, have been driven largely by piecemeal reaction to the manifestation of acute threats to human health or other anthropocentric concerns. Many examples from some of my other books highlight how externalities stemming from a lack of systemic selection can result in net erosion of the integrity and supportive capacities of ecosystems, distributional equity across societal sectors, and net value to humanity. We see this too in some environmentally based innovations. In the management of water, for example, the design, scheme evaluation, and approval of large dams solely on the basis of the value from piped water and hydropower still generally form the business case today. However, this myopia tends to overlook substantial disbenefits created by the drowning of valleys and displacement of wildlife and communities, often profound degradation of fluvial habitat-forming and soil regeneration processes, shifts in flow regimes favouring the spread of the vectors of waterborne diseases, methane generation from deep anoxic waters and a range of other negative ecosystem service consequences, in addition to long-term impacts such as dam siltation and the legacy risks of decommissioning.[3, 4] Likewise, the generality of intensive agricultural production tends to measure benefits in narrowly framed terms of financial returns from commodity production over and above inputs, whilst largely overlooking multiple interconnected consequences such as degradation of biodiversity, release of sequestrated carbon, soil erosion, nutrient cycling, valued landscapes, and wild species of conservation, aesthetic, recreational or potential genetic or medicinal use.[5] Neither are the full implications for the atmosphere considered in an integrated way in a fragmented policy environment that divorces control of climate-active, human health-related, acidifying, ozone-depleting, and other gases, resulting, for example, in trade-offs such as the generation of a greater volume of polluting substances when traffic is rerouted to overcome localised aerial pollutant build-up in urban centres.[6]

Consequently, the pace of environmental destruction wrought by cascading innovations built on narrowly conceived legacy assumptions, exacerbated by a booming human population and a changing climate, is far greater to date than our positive yet largely reactive responses. The broadening divergence between human pressures and the resilience of supporting ecosystems now makes it starkly clear that we lack the luxury of time or of an alternative 'lifeboat' planetary ecosystem for us to continue to respond to acute pressures in a retrospective and largely undirected way as multiple, compounded 'natural selection' pressures come to bear on our expectations and prospects. Transition to a wilfully directed and orchestrated form of change is a pressing requirement, necessarily entailing the building of a far more multifactorial, systemically connected set of criteria to inform our 'artificial selection' to products and practices. Only through such a revolution of greater synergy and sustainable interaction with the non-substitutable ecosystems supporting our continuing wellbeing can continuing security, equity and collective prospects be founded. This requires collective purpose, consensus and transparent assessment to drive an ecosystems revolution, directed at a strategic level although still comprising and guiding iterative innovations selected as more or less fit for a consensually chosen future.

Both undirected evolutionary change and directed revolution obey the same law of selection of incremental innovations and other 'mutations'. However, these different types of regime change are distinguished to the extent that they account for their consequences for ecosystems, people, and net societal value. Reviewing prominent breakthroughs in the ascent of humanity against this distinction offers insights into the nature of our past progress. This, in turn, may illuminate necessary characteristics of future direction, and the shaping of a chosen future in which we stop inadvertently undermining the very natural assets underwriting future potential and instead integrate them as the most valuable of resources securing out continuing progress. We are, after all, in no ways immune from the blind justice of natural selection, as masterfully reviewed across the sweep of human history in Jared Diamond's 2005 book *Collapse: How Societies Choose to Fail or Succeed*.[7] It is a wiser strategy to opt instead to unleash our advanced knowledge and unique human capacities for innovation and foresight to found our artificial selection on a more systemic basis, recognising that nature's limits are ultimately non-negotiable, and that a pathway of development consistent with natural selection pressures is one that is in our best collective interests. Do we, in short, continue to select for short-term gain at unforeseen cost, or for long-term security? Furthermore, how do we share the benefits and costs across society and generations within our model of progress? And what do appropriate governance models, respecting cultural diversity both nationally and internationally, look like for a global population heading inexorably beyond 9.5 billion by

2050, supported by a currently fast-diminishing natural resource base? We need all our cumulative best intentions, innovation and faculty for collaboration to address a wider field of human and ecological concerns to bring about a sustainable ecosystems revolution, the alternative to which is continuing decline of health, opportunity, and competition for dwindling resources.

PRINCIPAL VECTORS OF A DIRECTED REVOLUTION

Most historic human progress has been consequent from essentially undirected, fortuitous innovations that have offered competitive advantage in overcoming immediate limitations. However, some of our revolutions have been distinctly directed. An oft-quoted example comes from the American Space Programme. In a frequent retelling, beloved of management consultants yet in reality probably apocryphal, President John F Kennedy noticed a janitor carrying a broom during a visit to the NASA space centre in 1962. He interrupted his tour, walked over to the man, and said, '*Hi, I'm Jack Kennedy. What are you doing?*' In response, the Janitor is said to have replied, '*Well, Mr President, I'm helping put a man on the moon*'. Whether a true account or not, the tale does exemplify a team ethos within which everyone, from the Director through to the technicians and right down to the janitor understood the importance of their contribution to achievement of a goal that demanded substantial collective ambition, innovation and action if it was to be achieved. As we observed in Chap. 3, achievement of a goal beyond the capabilities of the then current technologies spurred and accelerated innovations in integrated circuits, which in turn triggered an explosion in communications, data processing, globalised markets and trading, amongst a wide array of human endeavour. The same is true of the advances in propulsion, materials, and related technologies essential to achieve the bold ambition of putting a man on the moon, not to mention the development of team dynamics to bring the efforts of many different players into alignment. The directed goal of putting a man on the moon required clear vision and strong intent, taking account of systemic ramifications of each innovation for the whole project, but triggering in its wake a multitude of tangential revolutions in other walks of human life.

Other examples of directed revolutions include the post-Second World War 'Green Revolution' comprising cumulative innovations driven by the threat of food security and the bending of agriculture to a new commer-

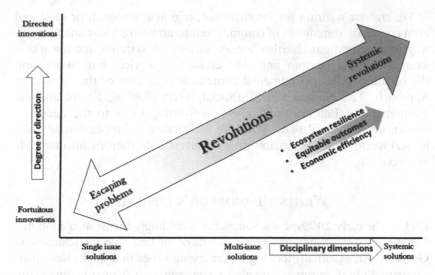

Fig. 4.2 Revolutions that are both directed and systemically assessed contribute to sustainable development, rather than merely reacting to immediate problems

cialised model, a clear goal with an associated set of 'artificial selection' criteria albeit omitting consideration of many consequences for ecosystems. Further directed revolutions include concerted medical campaigns to control or eradicate epidemics or diseases such as tuberculosis, polio, smallpox, HIV and Ebola. Also in the field of medical innovation, the Human Genome Project acknowledged at inception that it depended on innovation of isolation, replication, analytic, and interpretive technologies that were not available at the time the project was launched. All of these examples demonstrate how humans can imagine and mobilise to address an 'impossible' challenge, directing a revolution that brings with it unanticipated extensions of technical capability, novel applications and market opportunities.

Two principal vectors seem essential for driving this form of directed revolution: direction and dimension. When these vectors are combined, we move from simply retreating from today's problems and pressures, potentially inadvertently creating other problems to confront us tomorrow, through to directed step changes selected on the basis of their systemic impacts, with an associated set of net co-benefits for ecosystem resilience, equitable outcomes and economic efficiency as illustrated in Fig. 4.2.

We are not wanting for international, regional, national, or local and even corporate statements of commitment to attaining a sustainable pathway of development. Neither are we lacking in systemic frameworks—from the 'triple bottom line', the ecosystem services framework, and the broader socio-environmental contextual principles of the Ecosystem Approach, The Natural STEEP (Social, Technological, Environmental, Economic and Political) and others—informing us as to the necessary conditions of sustainability. What we are lacking is proportionate shifts in 'real world' practice. So how do we convert bold rhetoric into tangible progress?

WRITING TOMORROW'S HISTORY

Back in the early 2000s, I was somewhat unwillingly thrust into a job in a UK statutory environmental agency as one of the first 'horizon scanners' in Government, essentially looking for emerging issues that might have environmental ramifications. Being alert to emerging challenges is important, although I found the process fraught with difficulties. The lesser of these difficulties was trying to second-guess the inherently unpredictable trajectory of novel issues. Who, for example, could have foreseen the full pervasive and disruptive impacts of the internet, both positive and negative, or the occurrence or net impacts of the ash cloud event of 2010 as the Icelandic volcano *Eyjafjallajökull* erupted? But the major obstacle to being effective in that job was institutional, trying to incite interest and reaction from busy people in an organisation established around essentially siloed and fixed world views. Unfailingly, my presentations were met dismissively and with a frighteningly complacent '*We'll deal with that if and when it happens*' attitude. At one event, I was introduced as a 'Future Urologist' rather than a 'Futurologist'. The mistaken job description would have been, more or less, as effectual as I was in my Government agency role.

Profound and witty observations have been made about prediction. Niels Bohr, the Danish physicist who made foundational contributions to understanding atomic structure and quantum theory for which he received the Nobel Prize in Physics in 1922, famously observed that '*Prediction is very difficult, especially if it's about the future*'. To this, another famous physicist, Albert Einstein, added '*Occurrences in this domain are beyond the reach of exact prediction because of the variety of factors in operation, not because of any lack of order in nature*'. In essence, both were observing that the future is not necessarily entirely unknowable, but is subject to huge uncertainties pertaining to the many influences shaping it. Most predictions of the future simply get it wrong due to a failure to think systemically, instead fixating on narrow parameters and blinkering out confounding but potentially important factors that may exert profound influence on inherently complex and chaotic systems. For example, my boyhood magazines and many other more learned projections from the 1960s saw us all travelling around in hover cars and enjoying plenty of leisure time by the turn of the millennium, yet failing to anticipate today's problems of congestion, fuel shortages, and air pollution as we work ourselves into an early grave! Crude predictions or simple sets of scenarios, often the mainstay of policy formulation in government and business projections, then have limitations in shaping sustainable policy since outcomes will always diverge from forecasted projections. A different model is required to help direct decisions incrementally towards desirable societal outcomes.

The true nature of the world, and of today's problems, is that of complexity. The vexing challenges facing society were recognised in 1973 by Horst Rittel and Melvin Webber as constituting 'wicked problems', defined by being difficult or impossible to solve because of incomplete, contradictory and changing requirements that are often difficult to recognise.[8] The use of the term 'wicked' in this context recognises not that there is anything inherently sinister about these problems, but that they resist easy resolution due to complex interdependencies that may mean that a solution to one aspect may reveal or create other problems. Recognition of the 'wickedness' of problems explicitly embraces the systemic connections of today's pressing challenges, and hence the need for multidimensional thinking often in the face of conflicting priorities and world views. Faced with this reality, there is in fact no single 'right solution' serving all interests. The priority instead is that all dimensions of the problem are factored into decision-making frameworks that will inevitably entail political judgements, perhaps achieving a 'least worst' outcome that is deemed optimal across a range of interlinked consequences. If we are to make progress with the ecosystems revolution, it is necessary to embrace the high degree of uncertainty inherent in all decision-making, basing selection decisions on a rather broader set of parameters and potential consequences than has formerly characterised governance processes.

It is here that the concept and practice of backcasting can make an important contribution. Rather than starting from a basis of projections from the present into the future, backcasting instead takes as its starting point a desirable future, from which one then works backwards to chart incremental decisions, innovations, and policies that can lead stepwise towards that rather different future.[9] The outcomes of forecasting and backcasting approaches can differ radically. For example, in my book *Breathing Space*,[10] I draw out examples of how decisions about phase-out of CFCs in the late 1980s, as well as management of urban air quality and a range of other problems, might have had different outcomes. In the case of the recognised need to curtail releases of CFCs to protect the stratospheric 'ozone layer', a forecasting approach to 'getting out of today's problem chemical' led to the phasing in of also troublesome HCFCs, effectively sinking investment into tomorrow's 'problem chemical', whereas a backcasting approach would have triggered earlier innovation of a net atmospherically benign alternative.

We can add to this a contemporary example of recognition that the 'dash to diesel' in road cars in the UK during the late 1990s was, with hindsight, a wrong decision founded on failing to backcast from a broader range of sustainability concerns. Encouragement of a transition from petrol to diesel fuel, promoted through price and other signals by the UK's former Department for Environment, Transport, and the Regions, was driven by the higher fuel efficiency of diesel at a time when the science and politics of climate change were prominent in the immediate post-Kyoto Protocol era. Subsequently, increasing scientific knowledge about the contribution of fine particulates in urban diesel emissions to increased premature deaths, disproportionately affecting poorer urban communities, has driven political responses seeking now to discourage diesel.[11] There is also increased realisation that diesel fuel efficiencies are realised only on long journeys, with no carbon advantage in urban settings. Had wider factors been built into a backcasting approach, they would have influenced a different decision, the former pro-diesel policy now recognised as a wrong decision by ministers both in the UK and France.[12] A further negative consequence of this evolving, more multifactorial policy response is that drivers feel betrayed by responding to government encouragement, yet subsequently finding themselves paying more to use diesel vehicles which may also now have a reduced resale value.[13]

The vision upon which backcasting commences does not seek to know the finest details of a future world, but is developed consensually on the basis of principles defining the desirable future state. In essence, the food security agenda driving the 'Green Revolution', the aim of control or eradication of diseases and the ambition of the Human Genome Project, as discussed above, provided compelling shared visions for backcasting of incremental innovations and steps leading towards fulfilment of aspirational principles. Under The Natural Step framework, a scientifically based model founded on natural principles governing biospheric cycling with associated necessary conditions of sustainability defined as four 'System Conditions', is used as a basis for building a vision of the future sustainable characteristics of a product, municipality, infrastructure project, societal use of a material or other challenge from which a 'future history' can then be plotted as a set of incremental steps leading from current situation to the longer-term goal of a fully sustainable outcome.[14] This generates a 'road map' for business people, planners, government employees, and others concerned with finding sustainable solutions, against which to innovate. They may do so individually or with partners or affected stakeholders, to develop solutions that offer immediate rewards (continued profitability is an important such immediate consideration for businesses just as good value is for other enterprises) yet which constitute steps recognised as leading towards a clearly articulated if often long-term sustainable end-goal. Without backcasting, it is all too easy to strand investment in blind alleys, for example locking into a return-on-investment period for energy or water efficiency measures implemented on processes on materials that themselves may be a higher priority for deselection when longer-term sustainability goals are considered.

One particularly significant example of backcasting undertaken using The Natural Step framework is that of the UK's PVC (the plastic polyvinyl chloride) manufacturing industry. By the late 1990s, non-governmental organisation (NGO) pressure in Europe had branded PVC a pariah material, facing the industry with threats of extinction across a number of countries. The defensive position—that many PVC products conferred benefits to society in terms of, for example, medical applications, energy efficiency and affordability—no longer deflected rising societal concerns about a range of other environmental and health concerns connected with the plastic. Leading players in the UK PVC industry decided that action had to be taken, taking the bold step of approaching the UK office of international sustainable charity The Natural Step (TNS), seeking answers to two questions: Could PVC be part of a sustainable future; and, if so, What needs to be done to achieve it? This initiated a research and stakeholder engagement process using The Natural Step framework, including backcasting, culminating in the identification of five major challenges that the industry would have to address to work towards a sustainable future. Leading sectors of the UK industry embraced these challenges wholeheartedly. This process and the details of analyses and associated innovations and issues are documented in my book *PVC: Reaching for Sustainability*.[15] The five TNS sustainability challenges for PVC have since shaped innovation and commitment by the industry at increasingly broader scales, today forming the basis of the *VinylPlus* voluntary commitment of players in the PVC industry, including its supply chain and many major users, across the whole EU.[16] Backcasting was undertaken at the level of sustainability principles (the four TNS System Conditions), with key areas for stepwise progress simplified as five sustainability challenges. These sustainability challenges served to promote engagement across the whole PVC industry and other sectors (waste management and recycling, building specification and construction companies, regulators, etc.), influencing the wider societal life cycle of PVC products. Companies are free to innovate and complete through evolutionary innovations that are aimed at working towards the longer-term directed and consensual goal of full sustainability for the societal use of the material. Artificial selection within the PVC industry is thus wilfully directed to the longer-term and commonly understood goal of sustainability, informed by the 'natural selection' criteria of scientific principles governing the workings of the biosphere as internalised in The Natural Step model. Steps made towards sustainability by the PVC industry have since been impressive, and are certainly bolder and more transparent than for any other plastic and indeed for potentially all other materials.

Of course, other sustainability-related frameworks can be used as a basis for backcasting. These might include, for example, the ecosystem services framework,[17] the twelve principles of the Ecosystem Approach[18] which set a broader socio-economic context for addressing ecosystem services, or other tools such as the 'five capitals' model.[19] But the important point here is that a focus on the future at the level of principles can guide innovations and help direct step changes that will not be obvious or likely when forecasting from today. Some may relate to social, resource or environmental issues that are not yet on corporate 'radars' or subject to regulatory concern, yet are likely to influence resource availability, product acceptability or other dimensions related to future conflicts with environmental limits. Innovations shaped by backcasting can thus pre-empt pressures likely to arise in future, avoiding stranded assets and the need to react to pressures at the point at which they emerge. It is this process of navigation from future aspirations that can best serve addressing today's wicked problems, making them tractable and serving to bring people together to innovate progressive solutions—collaboratively, competitively or through 'co-opetition' (combining efforts on issues of innovation and principle yet competing at product level)—working towards the directed goal.

We have the knowledge and tools to work towards a chosen future, rather than accept a fate shaped purely by happenchance or the pursuit of immediate, narrowly framed rewards. Our chosen future needs necessarily to be informed by a greater symbiotic relationship between the choices we make and the dynamics of the natural world that supports our collective future prospects. This challenge is urgent given the burgeoning of human demands and the continuing decline and instability of those all-important ecosystems.

Lightening our collective load on the natural world is important, but is in practice an insufficient aspiration. Against the powerful intent and commitment to intergenerational equity of the 1987 Brundtland Report, *Our Common Future*,[20] we are profoundly failing in terms of leaving a global ecosystem that is anything short of seriously impaired. In large measure, this is because we have not responded proportionately to the challenges but also, significantly, there has been an underpinning assumption of 'stationarity'. Yet, as Milly and colleagues wrote with respect to water management systems, '*Stationarity is dead and should no longer serve as a central default assumption in water-resource risk assessment and management*'.[21] The harsh reality is that the baseline of ecosystem health, along with its supportive services, is in steep decline, including a destabilised climate, declining biodiversity and spiralling pressures from burgeoning human numbers and shifts in demographics and geopolitics.

Our collective thinking then has to move up a gear, to think in terms not merely of lightening our load but on reinvestment in and rebuilding of the supportive capacities and resilience of the life support system. Yesterday's risk-based management paradigm founded on the assumption that probabilities lie within static bounds, is now manifesting as a major source of uncertainty for both our engineered infrastructure and its consequences for nature's supportive infrastructure.[22] Rather, we have to raise our ambition to rebuild the damaged biospheric 'mother ship', the vitality and capacities of which will define human potential and opportunity into an, as yet, unwritten future history.

NOTES

1. Dawkins, R. (1986). *The Blind Watchmaker: Why the Evidence of Evolution Reveals a Universe without Design*. New York: W.W. Norton & Company.
2. Everard, M. (2015). *Breathing Space: The Natural and Unnatural History of Air*. Zed Books, London.

3. Everard, M. (2013). *The Hydropolitics of Dams: Engineering or Ecosystems?* Zed Books, London.

4. Everard, M. and Kataria, G. (2010). *The proposed Pancheshwar Dam, India/Nepal: A preliminary ecosystem services assessment of likely outcomes. An IES research report.* The Institution of Environmental Sciences, London. (https://www.the-ies.org/resources/proposed-pancheshwar-dam, accessed 23 April 2015).

5. Everard, M. (2011). *Common Ground: The Sharing of Land and Landscapes for Sustainability.* Zed Books, London. 214 pp.

6. Everard, M. (2015). *Breathing Space: The Natural and Unnatural History of Air.* Zed Books, London.

7. Diamond, J. (2005). *Collapse: How Societies Choose to Fail or Succeed.* Viking Penguin.

8. Rittel, H.W. J. and Webber, M.M. (1973). Dilemmas in a General Theory of Planning. *Policy Sciences,* 4, pp. 155–169.

9. Holmberg, J. and Robèrt, K.-H. (2000). Backcasting from non-overlapping sustainability principles: a framework for strategic planning. *International Journal of Sustainable Development and World Ecology,* 74, pp. 291–308.

10. Everard, M. (2015). *Breathing Space: The Natural and Unnatural History of Air.* Zed Books, London.

11. EPA. (2015). Health Effects. US Environmental Protection Agency. (http://www3.epa.gov/region1/eco/diesel/health_effects.html, accessed 12 October 2015.)

12. Hanley, S. (2015). Support For Diesel Was "Wrong Decision" Says UK Minister. (http://gas2.org/2015/02/04/support-diesel-wrong-decision-says-uk-minister/, accessed 12 October 2015.)

13. Swinford, S. and Collins, N. (2014). Diesel car drivers 'betrayed' as EU cracks down on Britain over air pollution. *The Telegraph,* 1 August 2014. (http://www.telegraph.co.uk/motoring/news/11007326/Diesel-car-drivers-betrayed-as-EU-cracks-down-on-Britain-over-air-pollution.html, accessed 12 October 2015.)

14. Robèrt, K-II. (2008). *The Natural Step Story: Seeding a Quiet Revolution.* New Catalyst Books.

15. Everard, M. (2008). *PVC: Reaching for Sustainability.* IOM3 and The Natural Step, London. 269 pp.

16. VinylPlus. (undated). VinylPlus. (http://www.vinylplus.eu/, accessed 12 October 2015.)

17. Millennium Ecosystem Assessment. (2005). *Ecosystems and Human Well-Being*. Island Press.
18. Convention on Biological Diversity. (undated). *Ecosystem Approach.* (www.cbd.int/ecosystem, accessed 12 October 2015.)
19. Forum for the Future. (undated). The Five Capitals. (http://www. forumforthefuture.org/project/five-capitals/overview, accessed 12 October 2015.)
20. World Commission of Environment and Development. (1987). *Our Common Future*. Oxford University Press.
21. Milly, P.C.D., Betancourt, J., Falkenmark, M., Hirsch, R.M., Kundzewicz, Z.W., Lettenmaier, D.P. and Stouffer, R.J. (2008). Stationarity Is Dead: Whither Water Management? *Science*, 319(5863) pp. 573–574.
22. Moddemeyer, S. (2015). Sustainability is dead: long live sustainability. *Water21*, April 2015, pp. 12–15.

CHAPTER 5

Reanimating the Landscape

Abstract 'Reanimating the landscape' recounts inspiring community-based projects across the developing world where restoration of degraded landscapes has regenerated ecosystems and human livelihoods in positively reinforcing cycles. Parallels are drawn with emerging developed world approaches to restoration of catchment functioning for pollution control, water resource protection, and flood management in increasingly nature-based ways. Examples from across the world illuminate how ecosystem restoration is protecting and increasing human security, economic benefits and opportunity, highlighting the importance of investment in the natural infrastructure essential for securing human wellbeing. However, significant difficulties are inherent in navigating a transition to a broader, systemic paradigm of greater net societal benefit and security, threatening as it may appear to established reductive norms and their associated vested interests.

Keywords Regeneration • Sustainable development • Community-based management • Socio-ecological systems • Reanimation • Restoration • Carrying capacity • Ecosystem services

The uncomfortable and generally unrealised reality is that we live on a planet with severely damaged ecosystems, the booming human population placing ever greater pressures on nature's dwindling supportive capacities. This, for me, is why the common understanding of sustain-

© The Editor(s) (if applicable) and The Author(s) 2016
M. Everard, *The Ecosystems Revolution*,
DOI 10.1007/978-3-319-31658-1_5

able development no longer encompasses a bold enough vision, as we make stumbling headway seeking to lighten and ultimately neutralise the tread of our demands on natural resources and processes. The stark reality is that those processes and the ecosystems that generate them are in far from a steady state. Rather, as evidenced by authoritative studies such as the UN's 2005 global Millennium Ecosystem Assessment[1] and the UK's 2011 National Ecosystem Assessment,[2] the baseline of nature is in steep decline, and with it human prospects to live secure and fulfilled lives.

As any businessman can tell you, seeking neutrality against a declining baseline is no formula for a sustainable outcome (or income). Instead, we have to elevate our vision to the rehabilitation of ecosystems, the fundamental natural infrastructure upon which humanity's prospects depend. Though the prognosis of the general cumulative momentum of humanity remains far from bright, there are, however, not merely green shoots but sturdy saplings of progress with ecosystem restoration serving as a basis for sustainable socio-economic regeneration.

Rehydrating the Desert Fringe

During one of my research visits to the desert edge of Rajasthan, I was particularly impressed by the fresh pugmark of a striped hyena in the soft earth of a newly constructed johad adjacent to Sariska National Park. Nearby, more established johadi (the plural of 'johad') and open wells were regularly visited by leopard and other species, even the occasional tiger, in search of a drink in the desert fringe.

A johad is basically an earthen bund intercepting monsoon rainfall, enabling water to percolate to ground and recharge aquifers. Johadi are actually extremely heterogeneous, some designed to hold standing water for livestock throughout the dry ten months, others also used for water-gathering or washing and bathing, and some with ghats (temple steps) giving worshipers access to the water's edge. However, recharge of groundwater in this formerly severely degraded semi-arid landscape is pivotal to the role of johadi in linked socio-ecological regeneration within a region of sparse rainfall and high evaporation.

Two days after seeing the hyena's pugmark, I renewed my acquaintance with Rudmal Meena, the headman of Harmeerpur village straddling the Arvari River. We talked as Rudmal Meena showed me around his wonderful garden. In the understory of trees—teak for timber and others for fruit including papaya, lychee, and banana—the garden was rich in orange, apricot, and almond bushes as well as a diversity of salad and vegetable crops and vivid hibiscus flowers. The garden also retained a cooler, shaded microclimate much appreciated not only by me but also by varieties of butterflies and birds.

Perhaps the most remarkable thing about Rudmal Meena's garden, and the rich surrounding village farmlands growing wheat, chana (chick peas), mustard, bindi (also known as okra or lady's fingers), onions, and many more crops vital for subsistence but also now used for trade, is that just 30 years previously this land was a desert: no trees, not even any grass. Johadi have played, and contribute to play, a central role in this remarkable story of rejuvenation of a tightly linked socio-environmental system. This is a tale that should serve as a global exemplar of how it is possible to reverse long-running ecological degradation and its disastrous consequences for social and economic wellbeing.[3]

Across Alwar District on the desert fringe of North Rajasthan, the NGO Tarun Bharat Sangh (TBS) has worked tirelessly at community level since 1985 to support people in taking control of water management, and thereby their own dependent prospects. Key to this has been the reinstatement of long-abandoned johadi and the design and construction of literally hundreds more rainwater harvesting structures to regenerate aquifers. Prior to this, groundwater levels had retreated to more than 100 metres below ground in places due to increasingly powerful mechanical pumps, with corresponding declines in levels in wells upon which less affluent villagers and farmers depended for year-round water availability in a landscape lacking in perennial surface waters and with high evapotranspiration rates.[4]

Johadi are augmented by many anicuts: low-level check-dams across shallow valleys that hold back water, inundating and nourishing soils during the brief and increasingly unpredictable monsoon period, and recharging adjacent aquifers. This diversity of water-harvesting systems has brought water back not merely to the soils and wells, but also rejuvenated the wider ecosystem and farming economy. It has also resulted in seven river systems—the Arvari, Sarsa, Baghani, Jhardwali, and Palasari as well as upper reaches of the larger Sabi and Ruparel systems—now holding water perennially, supporting a diversity of wetland flora and fauna, and restoring many benefits formerly lost to local people. Parched, overgrazed, and eroded landscapes have seen soil, water, and vegetation recover, and with it the vitality of communities, as monsoon rainfall once again percolates to raise water levels and also to sweeten deeper groundwater once causing localised fluoride-related health problems.

The effect has been truly transformative, reversing a long-running cycle of aridification, desertification, and village abandonment. Today, Harmeerpur, Gopalpura, Beekhampura, Kalid, and hundreds more villages in the region where communities have taken control of the most fundamental environmental resources securing their futures are experiencing full schools, active community participation in village governance, and access to more diverse and healthy food for their own needs and to trade for valuable income. Hand pumps installed close to people's homes now tap into a raised water table typically close to the surface or declining to perhaps 10 or 15 feet (3–5 metres) below ground according to season. This in turn has benefitted women massively, freed from the drudgery of their traditional role as gatherers of often poor-quality water and fodder from remote sources that formerly might have occupied seven hours of each day. Women are thus now empowered to play significant roles in village governance, education, and Ayurvedic medicine.[5]

At the core of these successes has been collective mobilisation and governance at community scale. Villagers deliberate and agree on water needs, on how johadi, anicuts, check-dams, and some other water-harvesting techniques may best help them rejuvenate the ecosystems supporting these needs, about the relative contributions that people should make both financially and in terms of labour, how water should be fairly and sustainably allocated, and what practices are to be prohibited (such as the growing of water-intensive crops like rice, the cutting of trees, and overgrazing).[6] Stewardship of this closely linked environmental and social system extends well beyond the river valleys, focusing also on promotion of the recovery of trees on surrounding hilly lands which are now greener after years as barren rocky crags, and that now once again store and recycle water and support grazing at agreed densities.[7]

Another intriguing feature is that johadi are designed through a rich amalgam of traditional and local knowledge, not by top-down imposition or assumption of uniform 'solutions'. Local people know the literal lie of the land and the disparate needs that johadi can serve within it, and so collaborate to optimise location, size, construction methods and materials, and management regimes.[8] They are supported with access to some funding, much of it donor aid from various sources most significantly in the past including the Swedish International Development Agency (SIDA),[9] and also by relevant know-how from TBS. This assistance includes, for example, support for reinstitution of traditional village decision-making arrangements and also technical aspects such as how to align johadi with fractures in the underlying rock strata visible from inspection of local open wells in order to optimise aquifer recharge. TBS also insists that the funding it channels from external sources is matched by local people, typically 30–70%, further grounding the johadi in local ownership.

A further remarkable facet of this rejuvenation is that it is, at least technically, illegal. The centralisation to national and state levels of the ownership and management of water under the late colonial era and India's post-independence period has poorly served rural people in marginal regions, such as the Thar Desert and its margins. Deriving uniform regulatory regimes enabling diverse, locally effective solutions consistent with geographical, cultural, and traditional stewardship contexts is always going to be difficult in as huge and heterogeneous a state as Rajasthan, India's largest state at 342,239 km^2 (40% larger than the combined area of the UK) encompassing major rivers and cities, mountains, wetlands, deserts and the Gangetic Plain. However, the disempowerment of communities through former centralisation, dispossession, and the ensuing abandonment of traditional local governance and physical rainwater harvesting infrastructure, has led to the linked degradation of soil and water resources with often severe and systemic adverse consequences for ecosystems and people. The same is observed widely in semi-arid environments across India,[10] Africa, China, and beyond, including for example America's 'Dust Bowl' experience of the 1930s precipitating the largest migration in that nation's history.

Driving this landscape rejuvenation in Rajasthan through constant encouragement, education, and routing of predominantly international funding has been TBS. During my March 2015 research trip to TBS, visiting and analysing key sites in the area and meeting with dedicated people significantly including TBS's founder and chief driving force Rajendra Singh, news came through that Rajendra had been awarded the prestigious Stockholm Water Prize,[11] a prize widely regarded as the 'Nobel Prize for water'.[12] We celebrated this honour in the cool oasis of trees, shrubs, butterflies and birds, fertile soils and both resident and visiting wildlife. This is a grand honour indeed, and it was wonderful to be there to celebrate over meals of chapatti, dahl, and other vegetable dishes all grown within the grounds of the TBS ashram at Beekhampura on land that had been restored from barren desert.

Due to the tight focus of TBS on working at community scale, including the linkage of communities across catchments to optimise benefits arising from collective action and governance, there are many more potentially significant benefits from landscape regeneration that remain uncounted. Benefits yet to be quantified include carbon sequestration in regenerated soils, climate mitigation and adaptation, flood regulation and the support of river and groundwater flows, benefits extending far further afield down sub-catchments and into incrementally larger receiving river catchments such as the Banas, Sawa, and Banganga, and downstream to the Chambal and eventually the Yamuna and the lower Ganga basin. Creative dialogue is also required with Rajasthan's state government, as well as Indian national government, to explore how the regulatory regime and its implementation can be modified better to support the kind of decentralised management that has a clear track record of sustainable success in these harsh conditions.

This inspiring story from the villages and catchments of Alwar District is one of the rebirth of a formerly damaged and degrading ecosystem, enabling the rejuvenation of nature and its capacities to support the recovery of community and economic opportunities. In this regard, it is truly inspirational, demonstrating nature's capacity to recover and for its range of beneficial services to be regenerated, where people collaborate in restorative action and appropriately adapt livelihoods and governance systems.

This tale of rejuvenation from the desert edge of Rajasthan gives hope, and also provides an exemplar from which we can derive lessons of vital importance for the future wellbeing of all of humanity. It is relevant to how we manage farmland in other environments and continents, how we harvest from the sea, and how we design cities. It applies even to those of us in the already industrialised world, buffered by historic trading advantages and overly comfortable assumptions but who are, in reality, very far from immune from the consequences of continuing global ecosystem degradation.

REBUILDING LANDSCAPES TO SUPPORT A BETTER FUTURE

The TBS-driven example in the desert fringe does not stand alone. There are also inspiring tales of rejuvenation of linked socio-ecological systems across India and in other areas of the world.

In China's Northwest, two major World Bank programmes have been restoring the Loess Plateau, home to 50 million people, where centuries of over-use and overgrazing have eroded the land leading to a widespread negative spiral of ecological, social, and economic decline and serious poverty.[13] The two projects—the *Loess Plateau Watershed Rehabilitation Project* and the *Second Loess Plateau Watershed Rehabilitation Project*—seek to reverse what was one of the highest erosion rates seen anywhere in the world, though loss of the dry, powdery wind-blown soil that gives the plateau its name.

The goal is to return this poor part of China to a condition supporting sustainable agricultural production. To date, more than 2.5 million people in four of China's poorest provinces—Shanxi, Shaanxi, Gansu, and the Inner Mongolia Autonomous Region—have been lifted out of poverty as degraded environments have been revitalised. This has been achieved through the introduction of sustainable farming practices including zoned grazing and terraced agriculture to protect soil, water, and nutrients, a focus on the protection of fundamental natural resources through controls on formerly unconstrained grazing, and limitations on other environmentally damaging farming practices such as cultivation on slopes and uncontrolled fuel wood gathering. The projects have encouraged natural regeneration of grasslands, tree and shrub cover on previously cultivated slope-lands, doubling the coverage of perennial vegetation and also doubling farmers' incomes and allowing for diversified employment.

Further benefits include dramatic reductions in the sedimentation of waterways, estimated to have reduced inputs to the Yellow River by more than 100 million tons each year and also slowing the infilling of dams. Food supplies have also been secured, cutting the need for government to respond with emergency food aid. The ecological balance has been restored in concert with a change in agricultural production producing a wider range of high-value products and greater productivity promoted by the creation of conditions for sustainable soil and water conservation.

The First Loess Plateau project cost US$252 million, of which US$149 million was contributed by the International Development Association (IDA: the World Bank's fund for low-income countries for which China qualified at the time). The Second Loess Plateau project cost US$239 million, with an IDA contribution of US$50 million. Although these sums are sizeable, the physical and economic transformation of the Loess Plateau offers a clear demonstration of the scale of linked socio-environmental regeneration that can be achieved if appropriate ecosystem-based restoration is undertaken in degrading areas of the world, leading to sustainable outcomes with multiple wider co-benefits arising from close partnership with the government, good policies, technical support, and active consultation and participation of the people. The projects' principles have since been widely adopted and replicated, the World Bank estimating that as many as 20 million people have benefited from uptake of the approach throughout China.

Similar results have been achieved in formerly highly eroded and degraded areas of the Ethiopian Highlands. Remote sensing data indicates that natural forest cover in the South Central Rift Valley Region of Ethiopia declined from 16% in 1972 to 2.8% in 2000, representing a cleared area of 40,324 hectares, an annual loss of 1440 hectares, and a total loss of 82% of the 1972 forest cover, indicative of trends in a region much larger than the study area.[14] These losses were attributed to small-scale agriculture and commercial logging and farming, the main consequences of deforestation being habitat destruction and decline of water availability. With increasing poverty allied with population growth in the highlands, cumulative pressures arising from the search for subsistence income has driven a spiral of massive deforestation and decreasing habitat functioning.[15] In order to halt and reverse this trend of upland desertification and poverty, revegetation has been identified as a key requirement.[16] Various projects have been developed to prevent further soil erosion and drainage of the water table, including through the rehabilitation of forests and the establishment of 'closed areas', where grazing is forbidden or restricted.

One such initiative is *Forest Rehabilitation through Natural Regeneration in Tigray, Northern-Ethiopia*, a joint research project of the Katholieke Universiteit Leuven (Belgium) and Mekelle University (Ethiopia) funded by the Belgian Government.[17] Other efforts include a regreening programme to restore one-sixth of Ethiopia's land through the planting of trees and shrubs, with plans targeting a further 15 million hectares by 2030 as part of a process that is already transforming degraded and deforested land across Africa.[18] Measures such as replanting trees and shrubs to stabilise ravines and slopes in the Ethiopian uplands are reversing a trend of abandonment of formerly severely eroding and barren areas that were plaguing communities with floods, droughts, and soil loss and leading to a constant requirement for food aid. Today, Abrha Weatsbha in the Tigray region, as one example of targeted ecological restoration dating back to 15 years, is unrecognisable, with the environmental catastrophe averted through the planting of many millions of tree and bush seedlings. Wells that had formerly run dry are now naturally recharged, the soil is regenerating, fruit trees grow in the valleys and the hillsides are once again green.

This 'regreening' approach has yielded dramatic and surprisingly rapid results at relatively little cost to Ethiopian farming communities in target areas. Here, communities have worked together to exclude grazing from large areas of the most vulnerable land, replant trees, adopt agro-ecology methods that combine crops and trees on the same pieces of land, and undertake water conservation measures.

In 2014, Ethiopia committed to outscale these successful measures across roughly one-sixth of the country in a concerted effort to reduce soil erosion, increase food security and adapt to climate change in one of the most drought- and famine-prone areas of the world.[19] This declaration was made at the 2014 UN Climate Summit under the New York Declaration on Forests,[20] a non-binding global pledge endorsed by dozens of governments, 30 of the world's biggest companies and more than 50 influential civil society and indigenous organisations to restore 350 million hectares of deforested and degraded landscapes by 2030. The intent is to cut natural forest loss in half by 2020 and strive to end it by 2030, cutting between 4.5 and 8.8 billion tons of carbon remobilisation annually (approximating that currently emitted by the USA). Ethiopia's commitment to restore 15 million hectares of degraded and deforested land to productivity by 2025, more than the total land area of Bangladesh, positions Ethiopia as a global leader in the restoration movement, and should yield environmental, social, and economic benefits for communities throughout the country reversing costs of degradation that currently exceed US$1 billion annually. Chris Reij from the World Resources Institute is quoted as stating that *'The scale of restoration of degraded land in Tigray is possibly unmatched anywhere else in the world. The people ... may have moved more earth and stone ... to reshape the surface of their land than the Egyptians during thousands of years to build the pyramids'*.[21]

Commitments to landscape restoration by other nations under the New York Declaration on Forests include Uganda (2.5 million hectares), the Democratic Republic of the Congo (8 million hectares), Colombia (1 million hectares), Guatemala (1.2 million hectares) and Chile (100,000 hectares), with many other nations expected to follow with their own commitments in the run-up to the United Nations Framework Convention on Climate Change (UNFCCC) Climate Summit in Paris in December 2015, as restoration of degraded land is expected to qualify for carbon credits.

Progress with linked co-management of water, soil and nutrients, and ecosystems supporting socio-economic stability and progress is being made elsewhere across the world. This is particularly the case in drier regions of the developing world, where these linkages are most directly and keenly experienced.

Another striking example is that of Zephaniah Phiri Maseko (born 1927), from Zvishavane in south-central Zimbabwe in a dry region suffering linked land and socio-economic degradation. Over a number of decades, Phiri has become known globally, and widely consulted by others across southern Africa, for his innovations in indigenous perma-culture and drought-sensitive farming methods that have won him the nickname of 'The Water Harvester'[22] or more generally 'the man who farms water'.[23] Efforts on his eight-acre landholding have transformed a hitherto dry area, suffering from fragile soils and erratic rainfall, into a perennial wetland through the innovation and application of unortho-dox and labour-intensive water-harvesting techniques. These include his now famous 'Phiri Pits', dug along contour ridges to capture rain water, particularly where directed by low bunds, resulting in raising of the water table to ensure constant moisture for trees and crops including bananas, sugar cane, beans, wheat, green maize, vegetables and various fruit trees and ponds containing fish. Phiri's methods were shaped by trial and error rather than book-based learning, and were declared illegal as they departed from proscribed methods. Phiri languished in gaol for a number of years until the Magistrate visited the landholding in person to review the outcomes.

Zephaniah Phiri's visit to the UK on the invitation of a British organ-isation helped him secure funds to found one of Zimbabwe's first Non-Governmental Organisations (NGO), the Zvishavane Water Project established in 1987, constituting an umbrella body of organisations and people studying and adapting his methods to the specifics of their local conditions. People have come from all over the world to see and learn from Phiri's successes, with Phiri himself invited to lecture and partici-pate in water-harvesting workshops and seminars both across Africa and overseas.

The underlying principles in Zvishavane, not to mention the frequent opposition from officialdom, in many ways reflect methods and percep-tions underlying traditional agricultural wisdom as embedded in the johadi of north Rajasthan, zia pits of West Africa, and a diverse range of other indigenous water-harvesting methods from across the world as reviewed in Fred Pearce's book *Keepers of the Spring*,[24] my own *The Hydropolitics of Dams*,[25] Brad Lancaster's *Rainwater Harvesting for Drylands and Beyond*,[26] and IMAWESA's *100 Ways to Manage Water for Smallholder*

Agriculture in Eastern and Southern Africa.[27] The organisation Excellent Development[28] actively promotes methods in east Africa to regenerate the water table, microclimate, and the ecosystems that support the livelihoods and prospects of local people in dry lands particularly through installation of 'sand dams', based on similar methods and principles to the johadi of Rajasthan, and which lead their local promotors to describe themselves as 'oasis makers'. Other organisations too, such as the Advanced Centre for Water Resources Development and Management (ACWADAM),[29] seek to promote wider uptake and innovation of locally adapted indigenous methods for the linked sustainable benefits of fragile, generally arid or semi-arid ecosystems and the people whose livelihoods are closely tied to them.

Although it is often the case that local farmers have identified 'development windows' best adapted to local geography, culture and needs, a lack of funds for 'bottom-up' research with farmers, and the acknowledgement and inclusion of indigenous knowledge by donor agencies still presents barriers to wider pervasion of such locally attuned methods.[30] Locally adapted sustainable water and ecosystem management technologies of this type have particular importance in regions of the world that are most vulnerable to climate change, such as various parts of Africa where extreme poverty, low adaptive capability and heavy dependence on rainfall renders people vulnerable to frequent natural disasters such as droughts and floods.[31, 32] Learning from local adaptation to climate change in these most vulnerable regions has further significance for wider global policy, as people in the Sahel between the 1960s and the 1990s have already successfully adapted to changes in rainfall on a scale at least comparable to that of future climate change scenarios through building flexibly on existing local knowledge to protect the assets upon which small-scale farmers and herders depend.[33]

This regreening approach represents something of a 'quiet revolution' with benefits for whole linked social and ecological systems, reversing former cycles of degradation, yielding real benefits for the security of communities and helping both to mitigate and adapt to climate change including rising temperatures and declining rainfall in vulnerable areas. Regreening, therefore, represents a key part of the solution to these linked problems, as agriculture, forestry, and other land use changes account for nearly 25 % of greenhouse gas emissions globally.

Rather than framed as a uniform, technocentric approach, embedding an implicit developed world paradigm in which market-based profit is a driving force, there is instead a reassignment of priority to what best supports local needs and works synergistically with local geographic character and culture. This alternative world view, directly pertinent to the needs of the global majority, has strong resonance with *Satyagraha* (holding to truth in all political and life decisions including equality and justice) and the quest for sufficiency as a goal as espoused by Mohandas Gandhi, and also with the championing of locally appropriate technologies encapsulated in the phrase '*Small is Beautiful*' by Fritz Schumacher.[34]

This approach—a form of 'pervasive localism' rather than a common interpretation of globalisation that frames development as leading people towards a market model—has a far more exact sense of sustainability at its core, in helping people meet their needs without compromising the ability of future generations to meet their own needs. Indeed, adequately supported, it offers the promise of going beyond sustainability into regeneration and the enhancement of opportunity.

CATCHMENT REANIMATION

It is not, however, just in the developing world that landscape-scale regeneration of linked environmental and socio-economic systems has been occurring. My book *The Hydropolitics of Dams*[35] reviews the outcomes of extensive installation of Integrated Constructed Wetlands (ICWs) in the Anne Valley, County Waterford, Ireland, and the 'reanimation' of the valley ecologically, socially, and economically. In summary, and without replicating the review of broad underpinning literature and the wider beneficial uptake of ICWs elsewhere in Ireland, the 'reanimation' has reversed a declining situation wrought by a quarter-century of subsidised arterial drainage converting the Irish wetland landscape for agriculture, commercial forestry and peat harvesting. Multiple values provided by wetlands formerly lost in a blinkered view of short-term commodity production are now being actively recovered. Although the use of constructed wetlands is not new, the ICW concept has at its core the integration of multiple ecosystem service outcomes provided by natural wetland processes, taking a 'landscape fit' approach that harmonises the reinstatement of cascades of shallow, vegetated wetland cells within natural, aesthetic, and working landscapes. Their benefits include efficient waste processing, hydrological buffering, regeneration of flows in the waterways of the Anne Valley, provision of valued amenity areas and formerly lost landscapes, and the recovery of populations of species such as otters, brown trout, salmon, sea trout and eels, and the interception of silt and nutrient from point sources. Networks of ICWs in the Anne Valley support farm profitability, manage sewage from household to industrial unit and up to village scales, also providing leisure opportunities and regenerating the formerly much degraded ecosystem.

ICWs have also been adopted elsewhere in Ireland for primary purposes such as treatment of landfill leachate, management of hotel waste, and of diffuse inputs in an urban context. 'Reanimation' is a term now often used to describe the extensive and dramatic regeneration of much of the Anne Valley, retaining water as a valued asset rather than seeking to rapidly shed it, and reintegrating the multiple, beneficial functions of the formerly widespread wetlands that had been eliminated from the landscape at huge and largely uncounted ecological, social and economic costs. ICWs have become incorporated into Irish Government design guidance within the *Water Services Investment Programme 2010–2012*,[36] recognising their potential to reverse former declines in the ecosystem services of lost natural wetlands, representing natural and nature-friendly solutions beneficial to the environment, farming and domestic needs, flood relief and many other linked benefits. Whilst no panacea for all ills, ICWs represent a low-input, multi-service output approach with many wider potential applications in Ireland and elsewhere in the world.

This focus on using or emulating natural processes to achieve multiple simultaneous ecosystem service outcomes has been described as 'systemic solutions', defined as *'low-input technologies using natural processes to optimise benefits across the spectrum of ecosystem services and their beneficiaries'*.[37] Systemic solutions thereby contribute to sustainable development by averting unintended negative impacts and optimising benefits to all ecosystem service beneficiaries, increasing net economic value. Other systemic solutions recognised under the initial definition include a converging range of urban ecosystem-based technologies and washlands, each of which yield multiple benefits, contrasted with the use of reed beds simply as tertiary wastewater 'polishing systems' that use natural processes but with limited service benefits.

Consideration of urban ecosystem-based technologies as potential 'systemic solutions' recognises that well-planned, sustainable cities can drive economic development, increase wealth generation and enhance human wellbeing within smaller footprints and with a lower per capita resource use and emission generation than any other form of settlement,[38] with integrated management of water crucial to the success of urban development. A range of ecosystem service-based urban development solutions are finding favour, including, for example, green roofs, water-sensitive urban design, green walls, rain gardens, rainwater harvesting, and other forms of 'green infrastructure', all of which bring with them contributions to reducing management costs and improving the quality of the urban environment.[39] Consideration of outcomes across a spectrum of ecosystem services recognises increasing potential synergies and eventual convergence between initiatives such as urban river restoration, natural flood management, sustainable drainage systems (SuDS), green infrastructure, Community Forests and nature-based management of climate-active gas emissions and impacts.[40] This cross-service realisation of benefits based on the restoration of natural processes is seen in the restoration of the River Quaggy in South London, UK[41],[42] and also the Mayes Brook as part of wider regeneration of Mayesbrook Park in East London,[43] in both cases breaking the rivers out of formerly tightly engineered channels into reengineered meanders across regenerated urban parkland. These schemes also not only delivered a softer and greener approach to flood management, but also restored a diverse riparian and floodplain wetland system valued for landscape aesthetics, local climate regulation, health benefits, access, wildlife, and amenity.[44]

Akin to ecosystem-based, potentially multi-benefit urban solutions—albeit that many applications still focus on one or a few service outcomes and are not optimised for cross-service benefits—water management solutions exploiting ecosystem processes at catchment scale are finding increasing favour at least in the developed world. Catchment habitat management for the protection or restoration of water quality have also been extensively reviewed in *The Hydropolitics of Dams*.[45] Global exemplars delivering cost-effective water resource protection—with a range of co-benefits including fishery, biodiversity, ecotourism, and stabilisation of farm business incomes—include the Upstream Thinking programme in south west of England under which South West Water (SWW), the regional water utility, recirculates a proportion of customer revenues into subsidies for improved agricultural land use within catchments serving surface water abstraction points.[46] This targeted investment reduces the quantity of particulate, soluble, and microbial pollutants at source to restore raw water quality as a key element of SWW's long-term aim to reduce the chemicals, cost, and energy needed to produce water for potable supply. Upstream Thinking is more than a simple altruistic programme: 'hard' business benefits accrue to the core water supply business of SWW and its customers. Water Services Regulation Authority (OFWAT), the economic regulator of the water industry in England and Wales, accepts that Upstream Thinking represents a 65:1 benefit-to-cost ratio relative to downstream treatment of contaminated water, with additional unvalued co-benefits for the ecological quality of rivers and associated fisheries, wildlife, and ecotourism.[47]

Upstream Thinking is one of many examples of schemes shifting focus from downstream technical solutions towards the upstream protection of ecosystem processes safeguarding the quality and quantity of raw water supplies. The Sustainable Catchment Management Programme (SCaMP) represents another pioneering European example. SCaMP was instigated by the British multi-utility company United Utilities, the water service provider for the north west of England. United Utilities is a major landowner, holding 57,000 hectares (140,850 acres) of upland principally to protect the quality of water entering reservoirs and rivers but also supporting nationally significant habitats for animals and plants. SCaMP was developed in partnership with wildlife NGO the Royal Society for the Protection of Birds. The first phase, formally funded by water service income between 2005 and 2010, entailed working with tenant farmers to alter land management practices and agreements and also undertaking additional capital works to restore upland habitat in ways simultaneously beneficial for water production, scarce ecosystems and agricultural methods stabilising farm incomes. Reinvestment of water service charges into upland restoration represented a cost-effective alternative to increasing water treatment and potential water shortages lower in the catchment. SCaMP therefore presents a 'win–win–win' scenario for biodiversity, society, and economic performance.[48] Subsequent phases of the SCaMP project have addressed water capture areas not owned by the utility company, but where targeted subsidy and advice on land use nevertheless culminates in water quality and quantity outcomes beneficial to the water service company, its customers, biodiversity and wider connected outcomes.[49]

Perhaps the largest global 'payment for ecosystem service' scheme, representing a contract between urban water users and the disparate farming and other communities in rural catchments, is seen in the Cat/Del (Catskills and Delaware) catchments providing raw water to New York City. By January 1997, constituent parties from both rural communities and the city formalised a comprehensive Memorandum of Agreement to which the city committed funds of approximately $US350 million (£190 million) in addition to the costs of various other initiatives in a watershed protection programme costing approximately $US1.3 billion (£700 million). This sum is substantial, yet represents only a small fraction

of the averted capital and operating costs, let alone the wider environmental impacts, from installation and operation of traditional filtration
solutions at the downstream point of abstraction to safeguard New York
City's water supply. This truly partnership-based approach, linking rural
and urban stakeholders into a mutually-beneficial arrangement based on
the ecosystem services provided by catchment land, is key to maintaining
New York City's pristine water quality as well as the viability of farming
for the foreseeable future.

In South Africa, the Working for Water (WfW) programme has also
demonstrated the benefits of landscape-scale, community-based integrated
management. The quality and quantity of water running off South Africa's
arid and semi-arid landscape has been compromised not merely by land
use but also significantly by invasive alien plant species, many of which
generate far greater evaporative loss compared to native, drought-adapted
species. The most widely accepted estimate for South Africa is that invasive
alien species could use 17% of mean annual runoff if left to invade. WfW
is an innovative scheme established in 1995, providing jobs and future
employment-relevant training for the least advantaged in society to control problematic invasive plants for the enhancement of water security,
ecological integrity, the productive potential of land, and for poverty alle-
viation. Today, WfW is one of the biggest conservation programmes in
the world, with a proven track record of benefits associated with improved
water yields and additional ecosystem services.[50]

Management of flood impacts has also seen a shift in focus from localised 'defence' of assets at risk towards more systemic, catchment-wide
solutions, some of which begin to recognise ecosystem service co-benefits
from landscape-scale interventions.[51] This is coalescing around the concept
of 'natural flood management' (NFM), defined as the alteration, restoration, or use of landscape features as a novel way of reducing flood risk.[52]
Potential co-benefits are also noted including reducing erosion and benefitting water quality, carbon storage, and biodiversity, providing positive
effects that may sometimes be more valuable than the reduction in flood
risk. Collaboration between landowners and communities constitutes a
key part of the success of NFM. However, this generally depends on long-
term funding measures or incentives, and better use of local knowledge. A
significant obstacle to NFM is that the full spectrum of ecosystem service
co-benefits may not be recognised in benefit assessment under driving
legislation and policy instruments (in England this falls under the *Flood
and Water Management Act 2010* and Catchment Flood Management

Plans developed by the Environment Agency) and in associated economic incentives. This lag between bold sustainable aspirations and international commitments to take an Ecosystem Approach, now both decades old, and the practical policy drivers shaping day-to-day decisions highlights a serious lack of coherence between stated intent and operational reality, and hence a high if serially neglected priority for review and reform.[53]

Aquifer storage is another process using a natural ecosystem for human benefit. We have already seen aspects of this in the role of johadi and anicuts in Rajasthan, as indeed sand dams of East Africa. However, the approach has been developed at a more industrial scale in the USA and Europe. It is important to distinguish between Artificial Aquifer Recharge (AR) and Aquifer Storage and Recovery (ASR). Essentially, AR is the enhancement of natural groundwater supplies using man-made conveyances, such as infiltration basins or injection wells. Exemplars of this are seen in the case of TBS interventions in Rajasthan, but are seeing wider applications elsewhere including in the USA and parts of southern England to augment natural water sources as climate change and development pressures increase demand on these finite supplies. Injection wells are frequently the selected method of artificial recharge in areas where the existence of impermeable strata between the land surface and the aquifer makes recharge by surface infiltration impractical, or in areas where land area available for infiltration is limited. By contrast, ASR is a specific type of AR practiced to augment groundwater resources for recovery for future uses, and is therefore likely to produce a more limited set of ecosystem-wide benefits as there is a narrower focus on water resources rather than wider potential benefits. English water company Thames Water is experimenting with a borehole a quarter of a kilometre deep near Horton Kirby, Kent, to understand the potential for drinking water to be stored in a 'bubble' 250 metres underground also relieving pressure on watercourses that suffer from excessive abstraction from surface waters and shallower groundwater, emulating ASR (aquifer storage and recovery) techniques used in desert regions across the world.[54] In California, the Orange County Water District has completed a $US142 million expansion of its Groundwater Replenishment System (GWRS), raising by an additional 38.2 m² of wastewater treated by reverse osmosis annually to reach a target of 127 m² to supply water to households containing 850,000 people.[55] In addition to recharging the Orange County Groundwater basin for safe, evaporation-free storage for later reuse, the injection of treated water through a series of recharge wells also provides a barrier against seawater intrusion. Although costly, GWRS water can be as much as half the cost of imported water, and has the advantage of being locally controlled. Also, as in other major water transfer schemes globally, there is rarely comprehensive consideration of the full breadth of implications for people and the ecosystems that support them in regions from which water is transferred. Where multiple uses of aquifers occur and/or where it is necessary to inject or encourage percolation of fresh water to deter salt water intrusion into freshwater aquifers and to control land subsidence, as in many regions of the USA, tight regulatory guidelines are necessary, for example, as guided by the US Environment Protection Agency.[56]

There is growing awareness of the potential for ecosystem-based landscape and waterscape management, protecting or regenerating natural retention of water, soil, nutrients, and biodiversity for a range of human benefits. These include, for example, natural regeneration of fisheries as opposed to a historic approach more focused on artificially maximising stocks,[57] and also for food security. Leadership in landscape management to create capacity for sustainable business is also evident in some of the activities of more far-sighted parts of the corporate sector. For example, under an ambitious sustainable development strategy launched in December 2014, Diageo, a global leader in beverage alcohol, announced 20 new sustainability and responsibility targets to be achieved by 2020.[58] The targets span leadership in alcohol in society, building thriving communities, and reducing environmental impact, building on former achievements and aligned with the United Nations' Sustainable Development Goals related to the areas of greatest material impact to Diageo's business. Under the 'reducing environmental impact' target, Diageo looks to its 130 manufacturing sites in over 30 countries, seeking better management of water stewardship and carbon emissions across their entire supply chains. Diageo's *Water Blueprint*[59] specifically includes a water stewardship strategy to support the company's expansion in emerging markets, not merely committing to cutting water use in half and returning waste water back to the environment safely across entire supply chains but, significantly, also replenishing water resources in water-stressed areas to an equivalent quantity used in final products through projects such as reforestation, wetland recovery, and improved farming techniques. Ownership of the entire water footprint is ambitious, the intent to regenerate water resources as a means to restore capacity for more sustainable operation being particularly visionary. Diageo does not stand alone, for example with SABMiller having helped fund four water recharge dams located on natural fissures in Rajasthan, retaining monsoon rains long enough to replenish aquifers as a means to regenerate groundwater resources depleted by its operations in India.[60]

A range of other business-led approaches to conserving the vitality and capacity of productive terrestrial and aquatic ecosystems and associated indigenous communities for a mixture of resource-stabilising and altruistic purposes are elaborated in Chap. 6. These approaches are novel and welcome, having at their heart the restoration of elements of ecosystem capacity supporting many formerly underappreciated benefits including resource security and wider human wellbeing.

RESTORATION ECOLOGY

The concept of restoration ecology is not new, emerging as a separate field of ecological science in the 1980s to support the practice of ecological restoration. There is a long-established Society for Ecological Restoration (SER), which publishes an online *International Primer on Ecological Restoration*.[61] The SER defines restoration ecology as '*the process of assisting the recovery of an ecosystem that has been degraded, damaged, or destroyed*'. A distinction is drawn between 'ecological restoration', the practice of restoring ecosystems by practitioners, and 'restoration ecology' that describes the science upon which practice is based.

As a generality, the starting point is an ecosystem in a degraded, damaged, transformed, or entirely destroyed state due to human activities, although possibly also aggravated by natural agencies such as wildfire, floods, storms, or volcanic eruption that take the ecosystem beyond a point from which it can recover its pre-disturbance trajectory. Typically, the reference for restoration is in some pre-disturbance state, for which long-term planning and commitment is then required and, often, consideration of links between the target area and the complex landscape in which it is situated. Restoration is complete when the ecosystem contains sufficient biotic and abiotic resources to resume its development trajectory without further assistance. The SER *Primer* outlines nine attributes for determining when restoration has been accomplished, including recovery of: characteristic species and community structure; indigenous species; functional groups necessary for continued development and/or stability; a physical environment sustaining core species; apparently normal functioning; integration with the larger ecological matrix or landscape; resilience against normal periodic stresses; and the ecosystem achieving self-sustaining properties.

In developing countries, many ecosystems are still managed through traditional, sustainable cultural practices, although these are under increasing pressure from demographic growth and a range of external pressures, including commercialisation of agriculture. Cultural landscapes or ecosystems are those that have developed under the joint influence of natural processes and human-imposed organisation, including for example many grasslands, savannahs, European species-rich meadows, and peat-dominated uplands generated by Bronze Age forest clearance. Although by their nature these systems are heavily human-influenced, they may constitute semi-natural baselines for restoration. Recovery of indigenous knowledge and ecological management practices therefore have key roles to play, although new culturally appropriate and sustainable practices that take into account contemporary conditions are often required.

A significant example of ecological restoration is seen in the Everglades system in the US state of Florida. The Everglades once constituted a vast, free-flowing river of grass extending from the Kissimmee chain of lakes to Florida Bay. This tightly connected series of wetland systems was rich in wildlife, ranging from dense flocks of wading and migratory birds, plants and fish, panthers, manatees and deer, and also conveyed clean water from Lake Okeechobee into Florida Bay and the coral reefs. Cumulative development impacts on the Everglades since the late 1800s, when canals were dug to begin draining South Florida, accelerated markedly throughout the twentieth century. The *Central and Southern Florida Project*, authorised in 1948, was a significant milestone, providing flood protection and fresh water to the burgeoning human population of South Florida, although achieved at high and largely uncounted cost to the ecology and ecosystem services of the Everglades. Today, more than half the Everglades wetlands has been lost to development, more than 1700 miles of canals and levees vastly changing the landscape[62] and reducing markedly populations of native birds and other wildlife, including the complete loss of some species. As a response to these evident gross cumulative impacts, the *Comprehensive Everglades Restoration Program* was initiated in 2000 by then-US President Bill Clinton[63] with the longer-term aim of restoring degraded swampland. The *Restoration Program* thereby seeks to overcome water quality problems through natural purification, boost biodiversity including a number of vulnerable species, add value to the tourism industry, restore natural floodwater controls, and serve a range of other associated benefits stemming from restoration of ecosystem services. The *Comprehensive Everglades Restoration Program* kick-started what has since become one of the world's largest natural capital restoration projects. Ecosystem restoration then is not merely 'for the birds', or at least solely for them; tangible and substantive societal benefits result as ecosystem functioning and formerly overlooked services are regenerated.

Similar, often surprising results have ensued when top predators, 'ecosystem engineers' and other key species have been reintroduced or allowed to recover as part of wider landscape restoration. One such example is the impact of the restoration of the grey wolf (*Canis lupus*) into several areas in the northern Rocky Mountains of the USA, including in Yellowstone National Park since 1995. Grey wolves had formerly been exterminated from almost all of continental USA, resulting in increases in populations of other animals, particularly large ungulates, with ensuing shifts in the character and functioning of the entire ecosystem of the American wilderness.[64] Beneficial influences arising from the reintroduced wolves almost immediately became apparent after they had been absent for nearly 70 years, with results continuing to unfold today. One of the initial impacts of the rising wolf population was control of numbers and an increase in overall health of the population of wapiti (a deer species: *Cervus canadensis*), but also changing ungulate behaviour. Wapiti spent far less time in the valleys and gorges, where wolves could more easily ambush them, and this promoted the re-establishment of native river corridor flora, increasing biodiversity, providing food and shelter for a growing variety of plants and animals but also, remarkably, changing the character of entire river systems. In particular, riverbank erosion decreased markedly, channels became less mobile and deepened, and small pools formed as recovering vegetation stabilised the banks. Wolves have thus proven essential to restoring the ecology of the wider Yellowstone region, including the many benefits flowing downstream to people through the waters of outflowing rivers. This phenomenon is known as 'trophic cascade', in this case where re-naturalisation of a top predator triggered profound effects throughout the entire ecosystem and the multiple, interconnected benefits it provides to humanity.

Where beavers have been introduced, they too have changed the dynamics of river systems. Beavers are, through their damming activities, known as 'ecosystem engineers' as they too have the capacity to change the characteristics of river systems. Across North America, North American beavers (*Castor canadensis*) were hunted to near extinction in the nineteenth century, whilst across Europe the range of the related Eurasian or European beaver (*Castor fiber*) has been severely curtailed including extinction from the UK. However, the role beavers can play in maintaining healthy river ecosystems is becoming better understood through a range of reintroduction programmes including, for example, the role of North American Beavers in enhancing stream habitat to the net benefit of endangered Chinook and Steelhead salmon in the Pacific Northwest in the Upper Columbia River Basin in Eastern Washington under the Methow Beaver Project.[65] Early results from the study of the reintroduction of Eurasian beavers into large enclosures in Devon, southern England, suggest that they may have a profound impact by increasing water storage and engineering woodland to release significantly cleaner water, which clearly has multiple benefits for ecosystems and the services they provide for people.[66] A review of outcomes from studies of reintroduction of beavers in Scotland[67] spread across multiple localised projects over 20 years support conclusions from other international studies a concerning the role of beavers as 'ecosystem engineers', their activities largely restricted to freshwater and associated riparian habitats in particular where broadleaved woodland is present leaving other habitats generally unaffected. Overall, beavers were found to have a very positive influence on biodiversity, creating new habitats and also increasing habitat diversity at catchment scale by creating ponds and wetlands, altering sediment transport processes, importing woody debris into aquatic environments, and creating important habitat and successional features. Whilst many species benefit from these changes, some others may be disadvantaged at local scales but may colonise new habitats created by beavers at wider landscape scales. Beaver activities enhanced a range of ecosystem services including increasing groundwater storage, regulating flows contributing to flood prevention, and enhancing people's recreational, educational, and spiritual interactions with the environment. Negative attitudes to beavers tend to be strongest where beaver activities affect intensive agriculture, and there may also be negative implications for some infrastructure such as weirs, roads, culverts, and fish passes.

Parallel conclusions have been drawn from bear re-introductions. The conservation of America's bear species depends on the conservation of the longleaf forests that support them, and so is associated with a wide range of linked societal benefits, just some of which include carbon storage, storm buffering, regulation of flooding, and the storage and purification of water. However, bears also play roles as predators and ecosystem engineers that maintain the diversity, functioning, and production of services. Other 'flagship' conservation species—charismatic apex species such as tigers, elephants, sturgeon, pandas, mahseer, and Atlantic salmon—can also mobilise support for protection and restoration of the networks of interconnected habitats they depend upon to complete their life cycles[68], [69] with associated uplift in other species and a linked range of ecosystem services beneficial to human communities.[70]

What has become starkly clear is the need to raise our vision from 'holding the line' against further habitat loss, instead working to create a world of rejuvenated ecosystems and the net reversal of declines in their quality, extent, and capacities to sustain continuing human security, wellbeing, and opportunity. As elaborated in Chap. 4, holding a declining line is now entirely insufficient an ambition for a safe and resilient future. Rather, we have to embrace the reality that planning for variability is as important as planning for constancy, emulating and seeking greater synergy with natural systems that have thrived through time by means of adaptive cycles, and thinking instead in terms of system resilience over and above planning for sustainability as if the world were static.[71] Crucially, this must entail investment in not merely protection but restoration of the natural infrastructure that is not only the source of future security and opportunity, but throughout epochs had ingrained within it by selective evolution a capacity for resilience and adaptability.

There is no greater investment in the future than upon the foundations of natural infrastructure. Commenting on the successes obtained by reinstitution of johadi and other water-harvesting structures and the village-scale institutions that support them in north Rajasthan, engineer G.D. Agrawal, commissioned to make an independent assessment of the work of TBS, concluded in 1996 that they not only stood the test of time but '*are, by and large, engineering-wise sound and appropriate*', concluding that '*There can be no better rural investment than on Johads*'.[72] There are, as we have seen, many positive examples of how this form of investment in natural infrastructure may be achieved, including both rural and urban ecosystems, and the range of societal benefits that can ensue. From these, we can therefore draw out key principles upon which to base a renewed ambition and relationship with nature and with each other.

ECO-CAPACITY

Rising human resource demands as global population heads towards 9.5 billion by 2050, the growing, resource-hungry middle class across the developing world exerting additional pressures adding to exacerbation by a changing climate, are leading to dramatic, systematic declines in the natural world. Various authoritative studies have quantified both the status of and the trends in this ecological decline, whilst addressing implications for continuing human wellbeing. These studies include the Millennium Ecosystem Assessment,[73] The Economics of Ecosystems and Biodiversity (TEEB),[74] 'ecological footprint' studies that standardise measures of human demand and waste assimilation relative to biologically productive land and sea area, and the United Nations Development Programme's Human Development Index.[75]

In 2009, the UK instigated its own National Ecosystem Assessment (UK NEA),[76] publishing a set of reports in 2011 that remain the only such national assessment globally. This led directly to a UK NEA Follow-on programme,[77] published in 2014, to address some knowledge gaps but, above all, to help communicate and spur action informed by the dense content of the first phase of the UK NEA. Implicit in both phases of the UK NEA was not only the development but also the transfer of knowledge, particularly to bring into the mainstream awareness and action to help reverse declining trends in ecosystems and their services as a contribution to resetting development on a more sustainable course.

Habitat degradation, over-exploitation, invasive alien species, pollution, and climate change are prime factors amongst a range of pressures affecting ecosystems across the globe.[78] About 60% of the world's ecosystems are degraded or used unsustainably, for example, with 75% of fish stocks over-exploited or significantly depleted and 13 million hectares of tropical forests cleared each year.[79, 80] Loss of biodiversity is proceeding at such a rate that we may face a mass extinction event if trends continue.[81] This decline in biodiversity represents not only an irreversible loss to the planet but also threatens humanity's life support system: the services that nature provides represent everything from the food we eat to the air we breathe, and primary resources of economic, spiritual, and a diversity of other values.[82, 83, 84]

The prime thrust of sustainable development commitments, particularly since 1992 when the concept broke into political consciousness following the Rio de Janeiro 'Earth Summit', has been to lighten the footfall of societal and economic lifestyles on supportive ecosystems. This is somewhat short of the bold vision of the 1987 World Commission on Environment and Development[85] that environmental, social, and economic progress should occur simultaneously. In the light of collision of sharply declining ecosystems, increasing demands from a growing and increasingly resource-hungry human population, and a changing climate, it has become essential to elevate the vision to one of not merely lightening our footfall but of rebuilding degraded natural carrying capacity if humanity is not to constrain its future potential.

Through the setting up of nature reserves, largely in response to emerging concerns about the loss of wildlife at the hand of other forms of human 'progress', we have at least made a start. For example, in the UK, the setting up of National Parks and other forms of nature reserve was established under the *National Parks and Access to the Countryside Act 1949*, amended by the *Wildlife and Countryside Act 1981* to strengthen protection of Sites of Special Scientific Interest (SSSIs). In the USA, the signing of the *Wilderness Act* in 1964 is another amongst a range of global examples of initiatives to permanently protect habitats for their inherent sake. However, our conception of the interlinked fate of ecosystems and human wellbeing needs radically to expand. The historic focus on designated species and habitats, and the establishment of nature reserves, has served society well throughout the past decades, and remains important for the protection of the most vulnerable aspects of biodiversity and geodiversity. Today, and increasingly tomorrow, pressures from surging human demands mean that this approach is now wholly inadequate to enable nature to adapt to changing conditions and to continue to safeguard the resilient provision of beneficial ecosystem services from wider non-designated landscapes, waterscapes, and the atmosphere.

Progressive thinking about conservation recognises the need to 'make space for nature' within broad economic landscapes, not merely in increasingly fragmented reserves. This was summarised in Professor John Lawton's report to the UK government on the future of nature conservation as a need for '*more, bigger, better and joined*'.[86] The vision of protection of nature has now to expand dramatically, both geographically and temporally, if ecosystems are to maintain their biological diversity, resilience and capacity to migrate with changing conditions. The necessary conceptual shift is from a paradigm of preservation for inherent worth, towards one of valuing and sustaining the supportive systems upon which continuing human wellbeing, prosperity, and security depend. And this shift in perception of value must, crucially, become increasingly internalised in the compulsions and inducements for land and other resource owners to 'farm' landscapes for the multiplicity of beneficial services best supporting continuing resilience and wider societal outcomes. For example, some 70% of the land area of the UK is farmed largely on the basis of rewards and sanctions related overwhelmingly to a single outcome—commodity production—softened only to a marginal extent by policy consideration of nature, heritage, and water management, just as mining, fishing, forestry, and other resource uses and ownership rights still largely favour exploitative practices that are substantially blind to their wider linked societal and ecological ramifications. Were the future defined as a simple race to the line—particularly taking account of the comprehensive current lack of balanced consideration of wider ecosystem service outcomes from the broad formal and informal policy environment of environmental and other statutory legislation, common law, catchment management strategies of varying influence, as well as subsidies, taxes and by-laws—it is very far from certain that extrapolation of the pace of the ecosystems revolution to date is even close to sufficient to overtake the momentum of legacy unsustainable practices.

Securing tomorrow's world, or at least a world where a decent quality of life is possible for humanity, depends upon us making more concerted efforts to accelerate the ecosystems revolution. It has been clear to science and environmental communities for many years that, if natural infrastructure continues to collapse or decline significantly in supportive capacities, the edifice of humanity implodes with it. Consequences extend from basic health and nutritional considerations to depletion of vital economic resources, and impoverishment of expectation of living culturally enriched and fulfilled lives. With declining natural capacity, we can expect greater conflict over dwindling resources. This is not merely the stuff of paranoia or science fiction archetypes, but an increasing reality as evidenced today by the extent of 'distress migration' from North Africa to Europe and between Asian nations, and in the water wars we have seen throughout history.[87] The fate of humanity is tied to that of the ecosystems with which we are indivisibly connected.

Dating back as far as the first half of the twentieth century, Mohandas Gandhi famously summed up the observed conflict of resource-hungry industrialised development with the quest for sufficiency for the mass of Indian and global poor with the statement that '*Earth provides enough to satisfy every man's need, but not every man's greed*'. More recently,

Edward O. Wilson, one of the world's leading champions of biodiversity protection and known as the father of sociobiology, crystallised much of his historic work in a 2013 book *The Social Conquest of Earth*,[88] which addressed three fundamental questions of religion, philosophy, and science. These were considered within a framework of evolutionary 'group selection', reigniting a long-running debate about how altruism evolved in humans, how morality, religion, and the creative arts are fundamentally biological in nature, and why the resultant human condition has culminated in our domination of the Earth's biosphere. Wilson espouses a new approach to conservation to avert what he sees as a coming 'biological holocaust': the planet's sixth mass extinction event driven almost entirely by human agencies.[89] Wilson calculated that the only way humanity can stave off a mass extinction crisis, one that spells severe limitation or irredeemable crisis for humanity, would be to set aside half the planet as permanently protected areas for the ten million other species. This 'Half Earth' goal would incorporate chains of uninterrupted corridors linking national biodiversity parks, which could address climate change goals by letting wildlife move south-to-north to accommodate warming and east-to-west in response to changes in rainfall. Whether the Half Earth vision is possible is perhaps a moot point; more pertinent is when humanity will respond proportionately to the fact that it cannot thrive without the processes of the living world, and hence the need to give them space not merely outside of the landscapes we use but also across and within them.

As long ago as 1992, the *World Scientists' Warning to Humanity* commenced with the statement that '*Human beings and the natural world are on a collision course*'.[90] This warning was not a petition of ill-informed amateurs but was led by Henry W. Kendall, former chair of the board of directors of the Union of Concerned Scientists, and was signed by a majority of living Nobel Prize laureates in the sciences as well as approximately 1700 of the world's leading scientists. It contained solidly science-based assessments about the collision course, including specific recommendations such as the pressing need to '*move away from fossil fuels to more benign, inexhaustible energy sources to cut greenhouse gas emissions and the pollution of our air and water*' and also that '*We must stabilize population*'. The Warning also highlighted that there was an ever-narrowing window wherein effective change should or indeed could be made without ensuing human misery. Little has changed since that time, other than a continuing decline in ecosystem capacity and the spiralling human population with its associated resource demands that conspire to accelerate the shrinkage of that window of opportunity.

We have, in short, to integrate our interdependence with nature into all facets of human enterprise. This is the nature of the ecosystems revolution. Some of the successes outlined in this chapter highlight that ecological restoration and an attendant expansion of eco-capacity to secure and sustain socio-economic opportunity and progress is not only possible but yields a host of benefits that have, in our recent economic history, been substantially overlooked.

MULTIPLE BENEFITS FROM NATURE

Despite this wealth of examples—drawn from rural and urban, developed and developing world settings—of how humans thrive when nature and its services are protected, restored or emulated, the bulk of societal activities have yet to be deeply influenced by the urgent need to reverse degradation of ecosystems and biodiversity. There are many examples of initiatives seeking to broaden the horizon by addressing additional ecosystem services, as for example research into maintaining an optimal flow of forest products under a carbon market.[91]

Globally, conversion and management for agriculture constitutes the greatest pressure driving the loss of terrestrial habitats and their associated ecosystem services.[92] Significant areas of global forest were found to be degraded, or subject to continuing degradation, with more than 14% of the area of tropical moist forest lost between 1950 and 1990 and more than 70% of temperate broadleaf forest and Mediterranean forests already lost by 1950 with commensurate losses in ecosystem services.[93] Forests are now recovering in some temperate countries, largely due to reforestation initiatives. Whilst plantations, particularly tree monocultures, do not have the complexity, biodiversity, attractiveness for amenity and wider suite of ecosystem services provided by natural forests, they can nevertheless play significant roles in the provision of non-timber ecosystem services when compared to agriculture and other forms of land use or where natural forests have been degraded.[94]

As with all ecosystem services, the quality, diversity, and location of the habitat with respect to natural and socio-economic landscapes play a significant role in the societal value of services provided by forests. For example, across the UK historic subsidies for monoculture plantations of coniferous trees in uplands have displaced peat bogs that formerly provided ecosystem functions valuable for carbon storage and climate regulation, water storage and purification, biodiversity and amenity.[95] By contrast, a case study under the UK National Ecosystem Assessment programme modelled the relative benefits and implementation costs incurred under contrasting 'market value' and 'social value' policies.[96] The study assumed that each country within Great Britain (England, Wales, Scotland and Northern Ireland) decides to plant 5000 hectares of new woodland per annum for each year between 2014 and 2063, yielding a potential increase in forest extent of 750,000 hectares.

The 'market value'-driven forest-planting policy scenario considered the situation if government were to seek to minimise the financial costs of meeting its afforestation targets, without taking account of the wider social consequences that planting trees might generate. Since forestry is invariably less profitable than the agriculture it displaces, this policy requires subsidies to be paid from the public purse to landowners in order to encourage them to plant trees. The 'best value' market-driven policy would seek to minimise the size of these subsidies after allowing for the value of any market-priced goods (predominantly timber) generated by the policy. Under this scenario, therefore, the distribution of forest planting is skewed towards the agriculturally less productive uplands. However, as observed above, uplands are important for many wider, generally historically overlooked and undervalued services such as water storage and purification, carbon storage and habitats for wildlife. Although the UK NEA study found that annual implementation costs were likely to be relatively low (£79 million) under the 'market-driven' scenario, there is a net negative return on investment (a net cost of £65 million) when consequences for these overlooked services are considered.

By contrast, modelling of the 'social value'-driven policy scenario addressed a wide range of social benefits in the location of forest planting, including both market-priced goods (such as the positive effect of timber production and the costs of displacement of agriculture) and selected non-marketed goods (such as greenhouse gas emissions and storage, and recreation). The need to pay subsidies is recognised, but policy emphasis shifts to obtaining the best social returns on investment in natural capital. Forest planting under the 'social value' policy scenario redistributes forest planting nearer to urban centres, closer to where people have access to the multiple benefits they provide. Annual implementation costs are relatively higher (£231 million), but there is a net positive return on investment (a net benefit of £546 million).

Comparison of likely outcomes under the 'market-driven' and 'social value' forest-planting scenarios reveals that, when the wider benefits provided by the natural environment are brought into decision-making, taking account of benefits across policy areas, rather different outcomes may ensue. In particular, 'social value' outcomes tend to both optimise public value and avert unintended negative outcomes across policy areas. This conclusion, mirrored in a range of additional case studies addressed in the same UK NEA study,[97] highlights the potential to generate net societal value when outcomes for a range of connected ecosystem services are considered.

This observation led the UK's The Natural Capital Committee (NCC[98]), in its third (2015) 'State of Natural Capital' report *Protecting and Improving Natural Capital for Prosperity and Wellbeing,*[99] to recognise 'natural capital deficits' built up over the long term that are proving costly to societal wellbeing and the economy. To counter this degrading trend, the NCC incorporated into its recommendations to UK government a 25-year plan that comprises, as one of three strategic themes, a strong economic case for investment in the creation and restoration of several optimally located habitat types. A strong economic case was presented for up to 250,000 additional hectares of woodland planting, peatland restoration on around 140,000 hectares of upland area, and wetland creation on around 100,000 hectares. Location is everything in terms of the production, access to and delivery of ecosystem services, with wetland creation delivering optimal value when located in areas of suitable hydrology, upstream of major towns and cities, and avoiding areas of high-grade agricultural land. The NCC study also made a strong economic case for commercial fish stock restoration, particularly white fish (like cod) and shellfish, the populations of which are today considerably below optimal levels. Intertidal habitat creation too could meet objectives set out in Shoreline Management Plans at substantial net societal benefit. High potential returns on investment through multiple ecosystem service benefits were also calculated as likely to arise from creation of urban greenspaces. The NCC found that returns on investment from this rebuilding of natural capital were at least as great as those from investment in traditional engineered infrastructure.

Remarkably, given the UK's fixation on economic growth as a dominating priority, and as if the economy inhabited some benign hyperspace with no environmental or social dependencies and consequences, a Natural Environment Bill began its passage through Parliament in Westminster in September 2015. The Bill's aim is to set biodiversity and other targets, re-establishing the Natural Capital Committee in some form, requiring local authorities to maintain local ecological network strategies, identifying species threatened with extinction, providing access to quality natural green space and including education about the natural environment in the curriculum for state-maintained schools. This was followed by a statement of commitment that *'The Government and interested parties endorse the Natural Capital Committee's proposed 25 year plan to maintain and improve England's natural capital within this generation'*.[100] Have we turned a corner, bringing the value of natural capital fully into market consideration and sectoral regulation? Only time and the judgement of future generations will tell.

MAJOR WORK TO BE DONE

Notwithstanding inspiring global exemplars and robust scientific studies of what is possible, much work needs to be done to convert the rhetoric of commitments to sustainable development into the reality of societal progress. And, as we have seen, this means not merely to slow and eventually halt the degradation of nature, but to rebuild damaged ecosystems for future security. The broader policy environment is still rife with embedded and often deeply entrenched societal expectations and assumptions as yet barely influenced by systemic thinking and the optimisation of outcomes,[101] much of it remaining rooted in industrial-era assumptions about the inexhaustibility of resources and externalisation of the wider costs of waste, resource depletion, and implications for non-consumers.[102]

It is timely that systemically connected approaches progressively supplant our patchwork of inherited, narrowly framed technical, legal and fiscal 'fixes', better to address emerging understanding of the 'wickedness' of today's challenges and to internalise the complexity and inherent interconnectivity of ecosystems and people. Navigating the ecosystems revolution means bringing government departments and their associated regulators, businesses and indeed all other sectors of society into a broader consensual vision. However, we know that paradigm changes are often perceived as threats to comfortabilities and power bases founded on legacy reductive, discipline-bound legislation, assumptions, models and economic instruments, not to mention often persuasive vested interests. It is indeed interesting to observe that larger potential benefits often generate greater resistance to change, perhaps because they span so many formerly narrowly defined fields of interest. But, if a revolution is built of evolutionary step changes, the task before us is to articulate the greater benefits of systemic change in ways that may be embraced as opportunities, much as the Stone Age and phonograph were laid aside because something better came along.

NOTES

1. Millennium Ecosystem Assessment. (2005). *Ecosystems and Human Well-Being*. Island Press.
2. UK National Ecosystem Assessment. (2011). *The UK National Ecosystem Assessment: Synthesis of the Key Findings*. UNEP-WCMC, Cambridge.
3. Everard, M. (2015). Community-based groundwater and ecosystem restoration in semi-arid north Rajasthan (1): socio-economic progress and lessons for groundwater-dependent areas. *Ecosystem Services*, 16, pp. 125–135.
4. Sinha, J., Sinha, M.K. and Adapa, U.R. (2013). *Flow—River Rejuvenation in India: Impact of Tarun Bharat Sangh's Work—SIDA Decentralised Evaluation 2013:28*. Swedish International Development Cooperation Agency, Stockholm. (http://www.sida.se/English/publications/Publication_database/Publications-by-year/2013/november/Flow-8211-River-Rejuvenation-in-India-Impact-of-Tarun-Bharat-Sangh8217s-work/, accessed 23 April 2015).
5. Singh, R. (2009). Community Driven Approach for Artificial Recharge –TBS Experience. *Bhu-Jal News Quarterly Journal*, 24(4), pp. 53–56.
6. Everard, M. (2015). Community-based groundwater and ecosystem restoration in semi-arid north Rajasthan (1): socio-economic progress and lessons for groundwater-dependent areas. *Ecosystem Services*, 16, pp. 125–135.
7. Singh, R. (2009). Community Driven Approach for Artificial Recharge –TBS Experience. *Bhu-Jal News Quarterly Journal*, 24(4), pp. 53–56.
8. Agrawal, G.D. (1996). *An engineer's evaluation of water conservation efforts of Tarun Bharat Sangh in 36 villages of Alwar District*. Tarun Bharat Sangh, Alwar.

9. Sinha, J., Sinha, M.K. and Adapa, U.R. (2013). *Flow—River Rejuvenation in India: Impact of Tarun Bharat Sangh's Work—SIDA Decentralised Evaluation 2013:28.* Swedish International Development Cooperation Agency, Stockholm. (http://www.sida.se/English/publications/Publication_database/Publications-by-year/2013/november/Flow-8211-River-Rejuvenation-in-India-Impact-of-Tarun-Bharat-Sangh8217s-work/, accessed 23 April 2015).

10. Kajisa, K., Palanisami, K. and Sakurai, T. (2004). *Declines in the collective management of tank irrigation and their impacts on income distribution and poverty in Tamil Nadu, India.* FASID Discussion Paper Series on International Development Strategies, No. 2004-08-005, Tokyo, Foundation for Advanced Studies on International Development.

11. The Hindu. (2015). Rajendra Singh wins Stockholm Water Prize. *The Hindu,* 22 March 2015.

12. Harrabin, R. (2015). 'Water man of India' Rajendra Singh bags top prize. *BBC News,* 21 March 2015. (http://www.bbc.co.uk/news/science-environment-32002306, accessed 23 April 2015).

13. World Bank. (2007). *Restoring China's Loess Plateau.* (http://www.worldbank.org/en/news/feature/2007/03/15/restoring-chinas-loess-plateau, accessed 7 April 2015).

14. Dessie, G. and Kleman, J. (2007). Pattern and Magnitude of Deforestation in the South Central Rift Valley Region of Ethiopia. *Mountain Research and Development,* 27(2), pp. 162–168.

15. Nyssen, J. (1997). Vegetation and soil erosion in Dega Tembien (Tigray, Ethiopia). *Bull. Jard. Bot. Nat. Belg./Bull. Nat. Plantentuin Belg.,* 66, pp. 39–62.

16. Chadhokar, P. and Abate, S. (1988). *Importance of revegetation in soil conservation in Ethiopia.* Pages 1203–1213. In: Rimwanich, editor. Constraints and solutions to application of conservation practices. Bangkok, Thailand.

17. *Forest Rehabilitation through Natural Regeneration in Tigray, Northern-Ethiopia* (VL.I.R. EI-2000/PRV-06), September 2000 to September 2004, Belgian Government Vlaamse Interuniversitaire Raad (VL.I.R.) fund.

18. Vidal, J. (2014). Regreening program to restore one-sixth of Ethiopia's land. The Guardian, Thursday 30 October 2014. (http://www.theguardian.com/environment/2014/oct/30/regreening-program-to-restore-land-across-one-sixth-of-ethiopia, accessed 8 April 2015).

19. Minnick, A., Woldemariam, T., Reij, C., Stolle, F., Landsberg, F. and Anderson, J. (2014). *Ethiopia Commits to Restore One-Sixth of its Land.* World Resources Institute, 21 October 2014. (http://www.wri.org/blog/2014/10/ethiopia-commits-restore-one-sixth-its-land, accessed 8 April 2015).

20. United Nations. (2014). *FORESTS New York Declaration on Forests Action Statements and Action Plans.* (http://www.un.org/climat-echange/summit/wp-content/uploads/sites/2/2014/09/FORESTS-New-York-Declaration-on-Forests.pdf, accessed 8 April 2015).

21. Vidal, J. (2014). Regreening program to restore one-sixth of Ethiopia's land. *The Guardian,* Thursday 30 October 2014. (http://www.theguardian.com/environment/2014/oct/30/regreening-program-to-restore-land-across-one-sixth-of-ethiopia, accessed 8 April 2015).

22. Witoshynshky, M. (2000). *The Water Harvester.* Weaver Press, Harare.

23. *Lancaster, B. (2008). CASE STUDY: drought resistant farming in Africa.* The Ecologist, *21 November 2008. (http://www.theecologist.org/campaigning/food_and_gardening/360257/case_study_drought_resistant_farming_in_africa.html, accessed 23 April 2015).*

24. Pearce, F. (2004). *Keepers of the Spring: Reclaiming Our Water in an Age of Globalization.* Island Press, Washington DC.

25. Everard, M. (2013). *The Hydropolitics of Dams: Engineering or Ecosystems?* Zed Books, London.

26. Lancaster, B. (undated). *Rainwater Harvesting for Drylands and Beyond.* (http://www.harvestingrainwater.com/resource-appendices/volume-1/# (Chapter, accessed 23 April 2015).

27. Mati, B.M. (2007). *100 Ways to Manage Water for Smallholder Agriculture in Eastern and Southern Africa: A Compendium of Technologies and Practices.* SWMnet Working Paper 13, IMAWESA (Improved Management in Eastern & Southern Africa), Nairobi.

28. Excellent Development. (undated). Pioneers of Sand Dams. (http://www.excellentdevelopment.com/home, accessed 23 April 2015).

29. ACWADAM. (undated). Advanced Centre for Water Resources Development and Management. (http://acwadam.org/home.asp, accessed 23 April 2015).

30. Maseko, Z.P., Scoones, I. and Wilson, K. (1988). Farmer-based research and extension. *ILEIA Newsletter,* 4(4). (http://www.agri-

culturesnetwork.org/magazines/global/enhancing-dryland-agriculture/farmer-based-research-and-extension, accessed 23 April 2015).

31. Watson, R.T., Zinyowera, M.C. and Moss, R.H. (2001). *IPCC Special Report on the Regional Impacts of Climate Change. An Assessment of Vulnerability.* Intergovernmental Panel on Climate Change.

32. Nkomo, J.C., Nyong, A.O. and Kilindwa, K. (2006). *The Impacts of Climate Change in Africa. The Stern Review on the Economics of Climate Change.* London: HM Treasury.

33. Mortimore, M. (2010). Adapting to drought in the Sahel: lessons for climate change. *WIREs Climate Change*, 1, 134–143.

34. Schumacher, E.F. (1973). *Small is Beautiful: A Study of Economics as if People Mattered.* Blond & Briggs.

35. Everard, M. (2013). *The Hydropolitics of Dams: Engineering or Ecosystems?* Zed Books, London.

36. Department of the Environment, Heritage and Local Government. (2010). *Integrated Constructed Wetlands: Guidance Document for Farmyard Soiled Water and Domestic Wastewater Applications.* 121 pp. (http://www.environ.ie/en/Environment/Water/WaterQuality/News/MainBody,24926,en.htm, accessed 3 June 2015).

37. Everard, M. and McInnes, R.J. (2013). Systemic solutions for multi-benefit water and environmental management. *The Science of the Total Environment*, 461–62. pp. 170–179.

38. Habitat UN. (2012). *Urban patterns for a green economy: working with nature.* Nairobi, Kenya: UNON, Publishing Services Section.

39. Grant, G. (2012). *Ecosystem services come to town: greening cities by working with nature.* Chichester, UK: Wiley-Blackwell.

40. Everard, M. and Moggridge, H.L. (2012). Rediscovering the value of urban rivers. *Urban Ecosystems*, 15(2), pp. 293–314.

41. Cowan, R., Hill, D. and Campbell, K. (2005). *Start with the park: a guide to creating sustainable urban green spaces in areas of housing growth and renewal.* London: CABE Space.

42. Potter, K. (2012). *Finding "space for water": crossing concrete policy thresholds in England.* In: Warner, J.F. and van Buuren, E.J. (eds.) Making space for the river: governance experiences with multifunctional river flood management in the USA and Europe. London, UK: IWA Publishing. pp. 89–102.

43. Everard, M., Shuker, L and Gurnell, A. (2011). *The Mayes Brook restoration in Mayesbrook Park, East London: an ecosystem services*

assessment. Environment Agency Evidence report. Environment Agency, Bristol.

44. Everard, M. (2012). *What have Rivers Ever Done for us?* Ecosystem Services and River Systems. In: Boon, P.J. and Raven, P.J. (eds.) River Conservation and Management, Wiley, Chichester. pp. 313–324.

45. Everard, M. (2013). *The Hydropolitics of Dams: Engineering or Ecosystems?* Zed Books, London.

46. Upstream Thinking. (Undated). Upstream thinking. (www.upstreamthinking.org, accessed 8 April 2015).

47. South West Water. (2012). *Corporate Sustainability Report 2012.* (https://www.southwestwater.co.uk/media/pdf/s/t/SWW-Corporate-Sustainability-Report-2012.pdf, accessed 8 April 2015).

48. Everard, M. (2009). *The Business of Biodiversity.* WIT Press, Lyndhurst.

49. United Utilites. (undated). SCaMP. (http://corporate.unitedutili-ties.com/cr-scamp.aspx, accessed 3 April 2015).

50. van Wilgen, B.W., Reyers, B., Le Maitre, D.C., Richardson, D.M. and Schonegevel, L. (2007). A biome-scale assessment of the impact of invasive alien plants on ecosystem services in South Africa. *Journal of Environmental Management* (doi:10.1016/j.jenvman.2007.06.015).

51. Everard, M., Bramley, M., Tatem, K., Appleby, T. and Watts, W. (2009). Flood management: from defence to sustainability. *Environmental Liability*, 2, pp. 35–49.

52. Parliamentary Office of Science and Technology. (2011). *Natural Flood Management.* POSTNOTE 396 (December 2011). The Parliamentary Office of Science and Technology, HM Government, London.

53. Everard, M., Dick, J., Kendall, H., Smith, R.I., Slee, R.W, Couldrick, L., Scott, M. and MacDonald, C. (2014). Improving coherence of ecosystem service provision between scales. *Ecosystem Services.* DOI: 10.1016/j.ecoser.2014.04.006.

54. WaterActive. (2015). Thames Water brings a little bit of Las Vegas to Kent. *WaterActive*, July 2015, p. 4.

55. Force, J. (2015). Orange County expands its groundwater replenishment system. *Water21*, June 2015, pp. 20–22.

56. US EPA. (undated). *Aquifer Recharge (AR) and Aquifer Storage & Recovery (ASR).* US Environmental Protection Agency. (http://

water.epa.gov/type/groundwater/uic/aquiferrecharge.cfm, accessed 8 April 2015).

57. Everard, M. (2014). Realising sustainable fisheries. *Fish*, 113 (Spring 2014), pp. 14–19.

58. Diageo. (2014). *Diageo commits to new ambitious sustainability and responsibility targets for 2020.* (http://www.diageo.com/en-row/newsmedia/pages/resource.aspx?resourceid=2411, accessed 8 August 2015).

59. Diageo. (2015). Diageo Water Blueprint. (http://www.diageo.com/en-row/NewsMedia/Pages/resource.aspx?resourceid=2730, accessed 7 September 2015).

60. Balch, O. (2012). Mouth Watering. Green Futures Special Edition—Water Works: Green Solutions for a Blue Planet. October 2012. pp. 22–23.

61. Society for Ecological Restoration International Science and Policy Working Group. (2004). *The SER International Primer on Ecological Restoration.* www.ser.org and Tucson: Society for Ecological Restoration International. (http://ser.org/resources/resources-detail-view/ser-international-primer-on-ecological-restoration, accessed 7 April 2015).

62. Everglades Restoration. (undated). Restoring America's Everglades. (http://www.evergladesrestoration.gov/, accessed 8 April 2015).

63. USGS. (2006). Comprehensive Everglades Restoration Plan (CERP). (http://fl.water.usgs.gov/CERP/cerp.html, accessed 8 April 2015).

64. The Guardian. (2014). *How Wolves Change Rivers—video.* The Guardian (GrrlScientist), 3 March 2014. (http://www.theguardian.com/science/grrlscientist/2014/mar/03/how-wolves-change-rivers, accessed 8 April 2015).

65. Methow Beaver Project. (2013). Methow Beaver Project: Accomplishments and Outcomes. Methow Beaver Project, April 2013. (http://www.methowrestorationcouncil.org/MethowBeaverProjectReport2013.pdf, accessed 8 April 2015).

66. Devon Wildlife Trust. (undated). *Devon beaver project.* (http://www.devonwildlifetrust.org/devon-beaver-project/, accessed 18 May 2015).

67. Scottish Natural Heritage. (2015). *Beavers in Scotland: A Report to the Scottish Government.* Scottish Natural Heritage. (http://www.

snh.org.uk/pdfs/publications/research/Beavers%20in%20 Scotland%20-%20Final%20-%2010%20June%202015.pdf, accessed 18 May 2015).

68. Caro, T. (2010). *Conservation by Proxy: Indicator, Umbrella, Keystone, Flagship, and Other Surrogate Species.* Island Press.

69. Everard, M. and Kataria, G. (2011). Recreational angling markets to advance the conservation of a reach of the Western Ramganga River. *Aquatic Conservation,* 21(1), pp. 101–108.

70. Everard, M., Fletcher, M., Powell, A. and Dobson, M. (2011). The feasibility of developing multi-taxa indicators for freshwater wetland systems. *Freshwater Reviews,* 4(1), pp. 1–19.

71. Holling, C.S. (1973). Resilience and stability of ecological systems. *Annual Review of Ecological Systems,* 4, pp. 1–23.

72. Agrawal, G.D. (1996). *An engineer's evaluation of water conservation efforts of Tarun Bharat Sangh in 36 villages of Alwar District.* Tarun Bharat Sangh, Alwar.

73. Millennium Ecosystem Assessment. (2005). Ecosystems and Human Well-Being. Island Press.

74. The Economics of Ecosystems and Biodiversity. (undated). The Economics of Ecosystems and Biodiversity. (undated). (www.teebweb.org, accessed 18 May 2015).

75. UNDP. (undated). *Human Development Index (HDI).* United Nations Development Programme. (http://hdr.undp.org/en/content/human-development-index-hdi, accessed 12 October 2015).

76. UK National Ecosystem Assessment. (2011). The UK National Ecosystem Assessment: Synthesis of the Key Findings. UNEP-WCMC, Cambridge.

77. UK National Ecosystem Assessment. (2014). The UK National Ecosystem Assessment Follow-on Phase: Synthesis of the Key Findings. UNEP-WCMC, Cambridge.

78. Pereira, H.M., Navarro, L.M. and Martins, I.S. (2012). Global Biodiversity Change: The Bad, the Good, and the Unknown. *Annual Review of Environment and Resources,* 37, pp. 25–50.

79. Millennium Ecosystem Assessment. (2005). *Ecosystems and Human Well-Being: Synthesis.* Island Press, Washington, DC.

80. UN FAO. (2011). *Payments for ecosystem services and food security.* United Nations Food and Agriculture Organization, Rome. 300 pp.

81. Barnosky, A.D., Matzke, N., Tomiya, S., Wogan, G.O.U., Swartz, B., Quental, T.B., Marshall, C., McGuire, J.L., Lindsey, E.L.,

Maguire, K.C., Mersey, B. and Ferrer, E. (2011). Has the Earth's sixth mass extinction already arrived? [online]. *Nature*, 471 (7336), pp. 51–57.

82. Díaz, S., Fargione, J., Chapin, F.S. and Tilman, D. (2006). Biodiversity loss threatens human well-being. *PLoS Biology*, 4(8), pp. 1300–1305.

83. Cardinale, B.J., Duffy, J.E., Gonzalez, A., Hooper, D.U., Perrings, C., Venail, P., Narwani, A., Mace, G.M., Tilman, D., Wardle, D., Kinzig, A.P., Daily, G.C., Loreau, M. and Grace, J.B. (2012). Biodiversity loss and its impact on humanity. *Nature*, 486(7401), pp. 0–9.

84. Hooper, D.U., Adair, E.C., Cardinale, B.J., Byrnes, J.E.K., Hungate, B. a., Matulich, K.L., Gonzalez, A., Duffy, J.E., Gamfeldt, L. and O'Connor, M.I. (2012). A global synthesis reveals biodiversity loss as a major driver of ecosystem change. *Nature*, 486 (7401), pp. 105–108.

85. World Commission of Environment and Development. (1987). *Our Common Future*. Oxford University Press.

86. Lawton, J. (2010). *Making Space for Nature: A review of England's Wildlife Sites and Ecological Network*. Department for Environment, Food and Rural Affairs, London.

87. Everard, M. (2013). *The Hydropolitics of Dams: Engineering or Ecosystems?* Zed Books, London.

88. Wilson E.O. (2013). *The Social Conquest of Earth*. Liveright Publishing Corporation, New York.

89. Hiss, T. (2014). Can the World Really Set Aside Half of the Planet for Wildlife? Smithsonian Magazine, September 2014. (http://www.smithsonianmag.com/science-nature/can-world-really-set-aside-half-planet-wildlife-180952379/#F24b5MSKhJEKU4GS.99, accessed 8 April 2015).

90. Union of Concerned Scientists. (1992). 1992 World Scientists' Warning to Humanity. Union of Concerned Scientists. (http://www.ucsusa.org/about/1992-world-scientists.html#.VFveWDSsWSo, accessed 6 November 2014).

91. Chikumbo, O. and Straka, T.J. (2012). Maintaining an Optimal Flow of Forest Products under a Carbon Market: Approximating a Pareto Set of Optimal Silvicultural Regimes for Eucalyptus fastigata. *Open Journal of Forestry*, 2(3), pp. 527–531.

92. Millennium Ecosystem Assessment. (2005a). *Ecosystems and Human Well-being: Synthesis*. Island Press, Washington DC.

93. Millennium Ecosystem Assessment. (2005b). *Ecosystems and Human Well-being: Current State and Trends, Volume 1.* Island Press, Washington DC.
94. Bauhus, J., Van der Meer, P. and Kanninen, M. (2010). *Ecosystem goods and services from plantation forests.* Earthscan, London.
95. Evans, C.D., Bonn, A., Holden, J., Reed, M.S., Evans, M.G., Worrall, F., Couwenberg, J. and Parnell, M. (2014). Relationships between anthropogenic pressures and ecosystem functions in UK blanket bogs: Linking process understanding to ecosystem service valuation. *Ecosystem Services,* 9, pp. 5–19.
96. Bateman, I.J., Day, B.H., Agarwala, M., *et al.* (2014). *Economic value of ecosystem services.* UK National Ecosystem Assessment Follow-on, Work Package 3.
97. Bateman, I.J., Day, B.H., Agarwala, M., *et al.* (2014). *Economic value of ecosystem services.* UK National Ecosystem Assessment Follow-on, Work Package 3.
98. Natural Capital Committee. (undated). Natural Capital Committee. (https://www.naturalcapitalcommittee.org/, accessed 8 April 2015).
99. Natural Capital Committee. (2015). *Protecting and Improving Natural Capital for Prosperity and Wellbeing: Third 'State of Natural Capital' report.* Natural Capital Committee, HM Government, London. (https://www.naturalcapitalcommittee.org/, accessed 8 April 2015).
100. HM Government. (2015). *The government's response to the Natural Capital Committee's third State of Natural Capital report, September 2015.* (https://www.gov.uk/government/uploads/system/uploads/attachment_data/file/462472/ncc-natural-capital-gov-response-2015.pdf, accessed 12 October 2015).
101. UK National Ecosystem Assessment. (2011). *The UK National Ecosystem Assessment: Synthesis of the Key Findings.* UNEP-WCMC, Cambridge.
102. Jackson, T. (2011). *Prosperity without growth: economics for a finite planet.* Earthscan, Abingdon.

CHAPTER 6

A Revolutionary Journey

Abstract 'A revolutionary journey' explores how an ecosystems revolution is already under way, as evidenced by incremental modifications to the broader formal and informal policy environment of the developed world over the past century and more. The dependencies and impacts of major policy areas on ecosystems and their services are reviewed through selected examples, emphasising the need for far greater internalisation of the benefits and vulnerabilities of supporting ecosystems, integrated across policy spheres and societal sectors, if continuing human opportunity is to be secured.

Keywords Benefits • Optimisation • Ecosystem services • Resilience • Revolution • Transition • Policy • Economics

Many truly inspiring examples in Chap. 5 highlight how an elevated emphasis on the vitality and functioning of ecosystems and their supportive capacities, instead of merely the narrow exploitation of just a few of their services, is already yielding significant benefits optimised across multiple ecosystem services. Thus, the resilience and functioning of the systems essential for future human security and progress are beginning to frame practice in at least some regions. The science base continues to grow in substance and extent further to propel an ecosystems revolution. Some progress is being made within policy areas too, as well as, importantly, also across them. These 'green shoots' of change highlight that a systemic

© The Editor(s) (if applicable) and The Author(s) 2016
M. Everard, *The Ecosystems Revolution*,
DOI 10.1007/978-3-319-31658-1_6

transformation—a revolution comprising sequential evolutionary shifts—
is possible given appropriate societal will and priority to overcome historic
myopia and current deeply vested interests. We also know this because we
have already taken surprisingly significant, albeit often disjointed, steps on
the journey.

A QUIET REVOLUTION

The transformational journey in the UK, mirrored across much of the
developed world, takes as its starting point the beginning of the twentieth
century when, as the common saying put it, '*An Englishman's home is his
castle*'. This statement neatly sums up how land, water bodies, and other
ecosystem resources were perceived largely as physical property, conferring
an associated set of substantially unconstrained rights to use that property as
the (generally male) owner desired.[1] However, by the close of the twentieth
century, the freedom of action of landowners was substantially constrained
by a corpus of environmental, industrial, planning, and other legislation, a
growing body of common case law relating to the impacts of resource use
on the rights of other people, incentives to manage the land in certain culturally preferred ways, taxes to dissuade undesirable activities, novel markets
for biofuels and feedstock crop production partly displacing dependence
on fossil resources, differentiated markets for sustainably sourced goods,
catchment management strategies favouring water-sensitive land uses, and
measures to secure public access, amongst a range of other changes. The
compressive effect of hindsight reveals what, in historical terms, was both a
profound and rapid transition, indeed a revolution of sorts albeit partial and
largely reactive to acute problems rather than directed.

Although this progression was hardly perceptible in those terms to
the people who lived through it, the net outcome has been transformative, constituting substantive early stages in a paradigm shift progressively reframing environmental assets not merely as private property
but as the source of a wide range of publicly beneficial ecosystem services warranting protection or promotion regardless of who owns the
productive resource. Beneficial services such as open spaces and fresh
air, buffering of storm energy, cleansing of air and protection of water
resources, conservation of soil and biodiversity, and heritage and landscape aesthetics have become progressively recognised and then subsequently valued and institutionalised, be that in statute or common law,
shifting social norms, market differentiation and reform, assumptions

embedded in models or accepted technical solutions, taxes and subsidies, or any of a range of other societal 'levers'.[2, 3]

So an ecosystems revolution is already in transit, albeit haltingly and in a fragmented, largely unrecognised and far from directed way with many services provided by this planet's ecosystems as yet poorly recognised and addressed. Nevertheless, the century-plus journey to date is already comparable in temporal scale and net impact to better-known 'revolutions', such as Europe's historic Agricultural and Industrial Revolutions. However, where the revolution required to escape today's diverse and pressing challenges differs is that we lack the luxury of centuries to make the scale of further transition necessary to secure a sustainable future.

Understanding the characteristics of the necessary transition then becomes a priority if we are to steer progress from evolutionary happenchance, with all the time lags and risks for ongoing system stability and extinction of opportunity that this implies, towards a more affirmative, directed revolution. Continuing and accelerating this transformational journey, addressing all spheres of human activities, is of crucial importance if we are to continue to live secure and fulfilled lives supported by the ecosystems with which we co-evolved.

REVIEW AND REFORM ACROSS SOCIETAL POLICY AREAS

As ecosystems and their myriad services underpin all dimensions of human wellbeing—provision of material substances and energy, regulation of environmental variables, cultural opportunity and enrichment, and supporting ecological processes that maintain the functioning and resilience of the natural world—it follows logically that all spheres of human activity are part of the systemic overhaul necessary to bring about the ecosystems revolution. As previously ascertained, progress has been occurring over the past century or so, although on a largely fragmented basis in retrospective response to acute problems as they have manifested. Much more is required, beyond bold rhetoric and ceremonial agreements at international and nation levels, to bring intention into the way we develop on an increasingly integrated and sustainable footing.

This section turns to key policy areas across society, as more or less reflected in the division of government departments across much of the world, as a principal filter. At least one theme is considered under each policy area, then looking at what integration across policy areas and their associated interests actually means, and how optimal outcomes can be achieved.

Treasury

As almost anyone who has worked in government or its agencies will freely tell you, there is in fact only one government department: the Treasury. The hand steering allocation of national finances is the arbiter of what can and cannot be done, at least from a top-down government perspective. After all, despite conflicting definitions of what exactly government is, there is at least consensus that it is a mechanism to control allocation of societal resources. So to understand the workings of Treasury departments, we have to look at the workings of the market.

The market, also by general consensus, is almost entirely dissociated from biophysical reality. We can go into a shop and buy an apple for a price, and this has been the nature of trading throughout most of human history. Yet, the proportion of global markets rooted in tangible goods today is vanishingly small. Instead, the overwhelming volume of trading is in things that do not exist, or at least do not yet exist. Stocks, anticipated future yields, swapping of debt in anticipation of high if risky returns, and so forth dominate transactions flashing at near-light speed across pan-global trading networks, exploiting second-by-second shifts in share prices to 'make money' with no connection to or liability for their consequences for the wellbeing of ecosystems and people. Yet tangible things, like apples, degrade over time. In a gravity-defying piece of trickery, interest on money borrowed to buy that apple, as indeed various forms of 'futures' markets, goes in the polar opposite direction, growing in time potentially beyond the capacities of the borrower to repay. Traders can sell on that risky debt, buyers playing an elaborate gambling game about the capacity of the borrower to honour interest payments that may spiral out of proportion to the meagre initial loan. And traders also speculate—another of those 'grown up' terms we use to disguise that it is in essence just gambling—about projected future yields of apples. It is not even as if they traded in 'real' money as, whilst governments make the notes and coins we use on a daily basis, in 2000 these constituted only 3.6% of the actual 'money' in circulation, the remaining 96.4% created by banks, building societies, and trading centres merely as computer entries[4]; this situation is sure to have compounded in the intervening years of progressively greater and more rapid digital trading. Thus, the virtual economy accelerates into a stratosphere almost entirely divorced from the entropic biophysical world inhabited by apples and people, including those in the less privileged underclasses who borrow to provide vital resources, such as apples, to feed their needy families.

The economy in turn drives the ways that everything from personal solvency to corporate performance and national wellbeing are measured, despite its dissociation from the natural world and the substantial externalities and associated risks from which traders are substantially insulated. The notion that all will prosper if the economy soars, a tenet of gross domestic product (GDP) and other metrics focused purely on gross economic activity, is fundamentally flawed. This is not merely in the demonstrably false assumption of a 'trickle-down effect', wherein the wealth generated by those at the top of the tree inevitably percolates to all. The absurdity of this assumption is exemplified by the 'horse and sparrow' metaphor, wherein if a horse eats enough oats then enough will pass through to feed the sparrows.[5] Inequities are also surely hard-wired if profit accrues to the affluent through lending their spare capital and trading in debts incurred by the least advantaged in society and between nations forced to borrow to meet their needs.[6, 7] Add to this repeated instances of whole economic edifices collapsing virtually in the blink of an eye, such as the folding of 'blue chip' institutions like Enron, Lehman Brothers Holdings Inc., Pan Am, or a host of other flopped mega-businesses formerly considered 'too big to fail'. It is also observed in the case of the sub-prime mortgage fiasco, rooted in the moral hazard of trading in unsustainable debts often foisted on those least able to repay, that triggered near-global financial melt-down in the early 2010s. This absurdity was illustrated in a thought experiment by Nobel physicist Frederick Soddy in the 1920s, conjecturing that, had Jesus invested £1 at the then current interest rate, the profit would have equalled the world's weight in gold; this skewing of wealth to the most privileged due to the capacity of virtual wealth to grow ad infinitum whilst physical assets decay, concluded Soddy, could only result in debt cancellation, revolution, or war.

Today, the wealthiest 1% of Americans possess 40% of the nation's wealth whilst the bottom 80% own just 7%, and the wealth of the 225 richest people in the world nearly tripled in the six years to 2000 with their assets equalling the entire annual income of half the world's population, the top 1% of households in the USA owning more wealth that the entire bottom 95%.[8] This magnitude of concentration of wealth by a privileged few at the expense of the increasingly marginalised many creates social instabilities, hyper-consumption by the affluent coincident with the poor overusing, and inadvertently degrading fundamental natural resources in a struggle to meet their basic needs. Market forces virtually wholly exclude most of the fundamental assets—the resilience and diversity of the natural world and its wealth of ecosystem services—that supports not only the long-term viability of the economy but of life itself. This contemporary market reality defies natural laws, at least during the limited period until natural limits and/or connected societal conflicts impose their own rebalancing through natural selection processes.

During the late 1960s, when some of the more acute adverse environmental effects of uncontrolled economic growth began to be appreciated, many economists began to question over-reliance by governments and other agencies on narrow, exclusively GDP-based measures of national progress. The principal flaw of GDP as a national or regional indicator is that it focuses purely on economic activity rather than its implications for people's wellbeing, perpetuating the primacy of the economy over all other facets of life including that of supporting ecosystems. This questioning prompted a search for a workable measure that took wider account of other dimensions of welfare. For example, the Measure of Economic Welfare (MEW) was developed in 1972, taking account of national output as a starting point but adjusting it to include assessment of the value of leisure time and the amount of unpaid work both of which increase the welfare value of GDP, but also of the costs of environmental damage caused by industrial production and consumption that reduce welfare value.[9] Successor metrics have built on principles underpinning the MEW approach to create increasingly sophisticated indices of sustainable development. For example, the Index of Sustainable Economic Welfare (ISEW) adjusts GDP further to take account of a wider range of harmful effects of economic growth, and by excluding the value of public expenditure on defence.[10] Other metrics include the Genuine Progress Indicator (GPI), which is broadly similar to ISEW with some models of GPI decreasing value when the poverty rate increases.[11] Other approaches include Gross National Happiness, a phrase coined in 1972 by Bhutan's fourth Dragon King, Jigme Singye Wangchuck, as a signal of commitment to building an economy that would serve Bhutan's culture, based on Buddhist spiritual values instead of western materialism.[12] The GPI was also updated in 2006 from a green measurement system to a broader concept that included quantitative measurement of wellbeing and happiness.[13] In 2011, the United Nations released the World Happiness Report,[14] informing a resolution passed at the UN General Assembly in July 2011 inviting member countries to measure the happiness of their people and to use this to help guide public policies. These and many more indicators including, for example, the Organisation for Economic Cooperation and Development's

(OECD) multifactorial Better Life Index,[15] have their champions and critics. However, none, to date, has challenged the supremacy of GDP despite its many obvious flaws in addressing sustainable and equitable progress, conspicuously including almost complete externalisation of the ecosystems vital for continued societal security and progress.

The overwhelming majority of the economy, in practice mainly an elaborate pan-global casino with the losers excluded from throwing the dice, continues to enforce supremacy by the privileged within and between nations. The fundamental question to be addressed in the ecosystems revolution is exactly who and what does the economy work for? What and who is included in the economy is an issue that matters very much, as an economy characterised by the bizarre oxymoron of excluding most of the fundamental natural and human resources underwriting its future progress must inevitably wither from the roots upwards.

Various commentators highlight how the use of money in modern capitalism has contributed to alienation, competition and scarcity, destroying community and driving endless growth for its own end. Charles Eisenstein posits a different sort of 'sacred economics', harking back to historic 'gift economies', recognising how money can instead be a driving force in the transition to a more connected, ecological, and sustainable way of being if outcomes are measured not in terms of personal wealth accumulation founded on little more than usury, but instead in societal enrichment and ecological resilience that each of us internally may hold most sacred and meaningful despite the everyday tyrannies of the market system we have created.[16] Eisenstein sees this evolutionary step towards a new economy as already under way, for example, in the form of new 'social currencies' of various online ratings systems that are, by their very nature, public and symbolic of trust and gratitude. This form of new currency is also connected to its 'story', quite unlike a Pound, Euro, Dollar, Rand, Rupee, or Yen that has no history, and therefore no such accountability for the processes that generated it. The loss of homogeneity of money, and hence a growing capacity to bear its history, may be facilitated by increasing digitisation of monetary exchanges which can carry with them an audit trail, potentially healing the formerly widening separation of profit from the consequences of investment. Eisenstein characterises this as transition to an 'Age of Reunion', the unfolding timeline of which is indeterminate, but in which it is possible to envisage an economy geared more closely to collective senses of beauty, equity, and ecological restoration. This may appear a utopian vision from current societal norms, but the alternative consequences of continuously divisive and destructive 'business as usual'—a trajectory that, let us recall, Frederick Soddy concluded, could only eventually culminate in debt cancellation, revolution, or war—provide their own compelling reasons for progressive decision-making founded on a different suite of societally beneficial outcomes.

In this connected world view, sense of self and the weight of feelings, from the pain of current trends to the aspirations and values that could frame a fairer and cleaner world, may and should progressively reform a system under which people have throughout history progressively allowed themselves to become subjugated.[17] A dominating market economy without conscience or reverence is a dangerous thing, creating profit in the absence of societal value or cognisance of its impacts on supportive ecosystems, and particularly given its increasing extension into developing regions. Yet the dangerous paradigms that people unconsciously allow to suppress their instincts and feelings, and to govern collective behaviour, are also within our collective gift progressively to change as it is we who are the actors in this 'system'. When one begins to think in these terms, established and emergent steps towards a new economy become evident beyond the 'social currencies' of ratings systems. Ethical investment banks and their products, concerned with the social and environmental impacts of investments and loans, represent a growing movement that, although regulated by the same authorities as traditional banks, commit to principles of transparency and accountability for the social and/or environmental aims of the projects they finance. Mainstream examples include the Co-operative Bank,[18] based in the UK, that only trades with businesses conforming to an ethical policy published since 1992 (although it should be noted that this bank was also a casualty of the near-global melt-down of the banking sector requiring a government bailout in December 2013), and the Triodos Bank[19] that undertakes 'positive screening' to lend only to businesses and charities judged to be of social or ecological benefit and (uniquely for a bank in the UK) providing an annual list of all the loans the bank has made.

Bonds for community energy, such as those supporting installation of solar farms or wind turbines, also serve as market instruments with clear pro-environmental outcomes. Microcredit schemes too, in which very small loans (microloans) are targeted at impoverished borrowers who typically lack collateral or a verifiable credit history, support entrepreneurship, contribute to helping people rise out of poverty, and provide financial services for those typically excluded from established banking models.

Microcredit schemes now represent a significant sector with an estimated 74 million men and women holding microloans in 2009 totalling US$38 billion.[20] So too the rise of crowdsourcing, in which investment or effort is solicited from a large group of people, especially through online communities rather than via traditional profit-taking enterprises, to promote services or innovations of various sorts.

An impressively system-challenging example here is the manner in which US President Barack Obama's first electoral campaign was substantially underwritten by 'grassroots' contributions from multiple 'small' individuals supporting the values he not only espoused but also published,[21] as an alternative to the established model of big corporate sponsorship. A similar 'line of sight' between investment and consequence is also seen in the rise of peer-to-peer lending, in practice often realised as a form of crowd-lending, that cuts out the banking middlemen, also often offering borrowers a slightly lower rate of return for longer-term commitment. The same potential is seen in the increasing use of crowdsourcing for ideas by businesses and to gather views by governments of all scales. It is also evident in the potential of pervasive digital media to elicit a greater diversity of perspectives to inform policy decisions which, if not automatically geared to optimally sustainable outcomes nor free from capture by powerful vested interests, certainly offer opportunity for greater democratisation and transparency in the shaping of consequences.

It is arguable also that traces of this values-based redirection of investment can be seen in the grand philanthropic gestures of the industrial era that saw the establishment of municipal medical, library, artistic, educational, and other institutions from which those in the industrial world still benefit today, albeit that these social improvement bequests were possible on the back of unprecedented wealth generated through practices with questionable impacts on ecosystems and people including the worst excesses of the age of empire. Arguably too, in the current era of super-rich minorities, we may be experiencing a new age of philanthropy, in which various individuals with way in excess of the money necessary to support themselves are making available substantial sums for altruistic purposes. The prime example here is the *Bill and Melissa Gates Foundation*,[22] the largest private foundation in the world, as of 31 March 2015 employing 1376 people with a Foundation Trust Endowment of $42.9 billion and having paid out US$33.5 billion in total grants since inception. The 'Gates Foundation' was launched in 2000 by Bill Gates, the Microsoft multi-billionaire, and his wife Melinda Gates with the primary aims of enhancing healthcare and reducing extreme poverty globally, and also expanding educational opportunities and access to information technology across America. The Foundation applies business techniques to giving through a 'venture philanthropy' model, which includes a willingness to experiment with novel approaches, focus on measurable results using mutually agreed metrics, fund on a multi-year basis and prioritise capacity building typically with a high degree of involvement between donors and their grantees. Such is the success of the Gates Foundation that, in 2006, Warren Buffett, then the world's richest person, pledged substantial sums from his fortunes spread over multiple years to be used as a matching contribution, doubling the Foundation's annual giving. Many large institutions too, from professional associations such as the UK's Royal Institution of Chartered Surveyors to major businesses such as the America-based Ford Motor Group, operate separate research and/or development Trusts. These Trusts are numerous and disparate, each with its own focal priorities, yet cumulatively represent a model of using money for variously framed environmental and social good, rather than for market-based return, therefore with some similarities to Charles Eisenstein's concept of 'sacred economics'.

Consideration of the workings of the market economy in this section is broad-ranging both in its extent and in consideration of the wider economy. Yet, as money is such a powerful influence on societal priority-setting, wider consideration of its current and potential relationship with environmental and social progress is as unavoidable as is the role of Treasury departments in setting presumptions and priorities for all other major societal policy areas. The direction of expenditure, investment, and rewards to restorative outcomes is a key feature of the ecosystems revolution. This book has observed examples of this tied to World Bank and other donor-led schemes, but also increasingly those tied more directly to business 'bottom lines' such as ethical banking and investment products, environmentally and socially oriented stewardship schemes, and catchment-based approaches to both water resource and natural flood management. So progress is evident across societal sectors, if as yet very far from a pervasive norm. Wholesale fiscal review and reform is thus an overdue element of the ecosystems revolutions, to reconnect the economic system more closely with its inevitably linked social and ecological consequences.

Business

Trade and industry are the engines of that flawed economy. However, they need not be, and indeed can become the principal drivers of an ecosystem-based revolution. Business, after all, is merely the mechanism by which capitalism, an ideology more globally pervasive than any faith,[23] converts natural resources into useful products through human labour and ingenuity at the heart of wealth creation.[24] As long as industry is compelled and rewarded by a diversity of societal rules and protocols to exclude externalised impacts on fundamental natural and human resources, it will continue to promulgate its vulnerability to poor risk management as well as liabilities that will increasingly impose themselves in an increasingly resource-constrained future.

Business has not to date taken full account of its dependency on water and other physical resources such as crops, stock, minerals, the productivity of the soil, biochemicals, and other natural stocks and flows, nor indeed of their vulnerabilities if the regulation of climate, water flows, soil erosion, pests, and diseases are compromised. In significant measure, this is due to the conceptual distancing of business managers buying these resources through intermediate suppliers, and the consequent perception that their dependencies are a commercial risk rather than related to stewardship of the basic resource itself.[25] Commercial pressures in the absence of an adequate policy environment therefore tend to underestimate or overlook interdependence between continuing profitability and ecosystem health.

If the purpose of business is to provide useful products to people, it is only right that full societal impacts—both directly or vectored via environmental media—are central to corporate governance and the wider inducements and compulsions of the policy environment that drives day-to-day decisions. One thing that business excels at is risk management. The thrust of my book *The Business of Biodiversity*[26] was to put business relationships with ecosystems and their services into the language of risk, terms with which business is highly experienced and capable of responding. Indeed, business has been in the forefront of consortia leading to development of the exemplar Forest Stewardship Council (FSC)[27] and Marine Stewardship Council (MSC)[28] initiatives, as well as a range of broader supply chain initiatives such as Rainforest Alliance,[29] Organic[30] and Fairtrade[31] brands, that seek to bring into the mainstream sound stewardship of natural resources and equity for indigenous people as a basis, amongst other outcomes, for securing stable resources into the future. Business too has been the driver of a range of other far-sighted, directed initiatives, such as the multifactorial quest for sustainability across the EU PVC industry, as elaborated in Chap. 4.

Business has often been castigated as at the root of raping ecosystems for short-term profit, and indeed has much blood on its hands on that score—some far more than others—albeit driven in decision-making largely by governance systems that reward immediate profit-taking that externalises most impacts on ecosystems and people. But business can be, and has not infrequently been, a leader in development of sustainable relationships with productive ecosystems if the need and benefits can be made evident and, ideally, progressively integrated into the mainstream of policy drivers.

Energy

Energy is a fundamental resource driving business and economic activity, and a principal need and cost for all societal sectors. Historically, energy policy has comprised exploitation of what was available and cheapest. It is, for example, far from coincidental that the birthplace of the Industrial Age was in the Ironbridge Gorge in the British West Midlands, blessed as it was with an abundance of readily accessible iron ore, water, and coal. Nor indeed that the furnaces of American heavy manufacturing in Detroit and Pittsburgh are well served by abundant fresh water from the Great Lakes and the Allegheny River, iron ore from the adjacent plains, and coal from the Appalachians.

Through ages of European empire-building, energy as well as a diversity of other raw resources has stimulated exploration, exploitation, and often annexation. Foreign policy too has been substantially shaped by domestic energy security, for example, with significant military interventions that secure sympathetic regimes in oil-rich states, whilst more flagrant atrocities across Eastern Europe and both west and central Africa have been allowed to continue without such avuncular concern.

Other factors have come to bear on energy sourcing, significantly including recent global concerns about the need to decarbonise the economy. Major progress has been made too by many industry sectors in energy 'footprint' contractions, linked as they are to resource costs and hence profitability. Progress towards decarbonisation targets too by those ratifying the Kyoto Protocol is welcome, albeit that the UK's 'dash for gas' as a less carbon-intensive fuel for generation of electricity was most strongly driven by regulatory changes allowing gas to be used for power generation, the lower cost of gas relative to alternative fuels, the speed at which gas turbine power plants could be built as compared to coal and nuclear power stations, innovation of combined-cycle gas turbine generators which provided higher relative efficiencies at lower capital costs compared to other conventional generation technologies, and the relatively recent development of North Sea gas; the contribution to Kyoto Protocol targets was almost entirely coincidental rather than deliberate.[32]

In terms of longer-term commitment to both decarbonisation of the economy as well as the phase out of nuclear generation, Germany's Energiewende[33] ('energy transformation') policy has positioned it as a global leader. By 2015, Germany had converted almost 30% of its electrical generating capacity to solar and wind power in just 15 years from virtually a zero baseline, supported by more than 1.4 million German householder and cooperative generation schemes.[34] Significant additional global contributions stemming from Germany's mainstreaming efforts include political and atmospheric security, also driving down cost performance including a price drop of 80% for solar and 55% for wind energy production through creation of mass markets. A powerful driver of this dramatic transition and a centrepiece of the Energiewende was a generous 'feed-in tariff' (FIT) rewarding home scale and community investment in solar and wind power systems in return for predictable financial returns. The high initial level of the tariff, paid for by a tax on energy use (albeit a policy not without its critics), served its two designed purposes of increasing installed renewable generation capacity and driving down costs to help these technologies achieve mainstream market penetration. The level of German FIT for new entrants has since declined, reflecting the reduced capital cost of scheme installation.

The UK too implemented its own FIT, replacing a former 50% capital grant in 2010. The UK FIT proceeded to drive early domestic installation with the promise of returns on investment well in excess of bank interest rates in the faltering economy, serving similar decarbonisation goals and innovation by business. About 28,608 installed photovoltaic systems, out of a total of 30,263 renewable installations (also including hydropower, wind, microCHP, and anaerobic digestion), were registered within the first financial year of the UK FIT scheme.[35] However, early in the life of the UK FIT scheme, one of many short-sighted reactions of Britain's Coalition government in the name of 'austerity measures' was successive significant reductions in subsidies for both home scale and large-scale solar arrays in both 2011 and 2012, slowing considerably investment in renewable energy options and the harvest of patents and other economic and environmental benefits that would ensue. How much more could have been achieved towards the wholesale replacement of problematic centralised generation technologies if the UK had shared Germany's vision and commitment? One has to question whether this 'knee-jerk' reaction is a triumph for financial prudence, or a spectacular failure to invest in the future security that a more stable climate offers, and the stimulus of new businesses better geared to service it, in deference to established vested interests.

Sweden is one of the most pioneering countries in the quest to achieve a sustainable future. Sweden's Prime Minister, Stefan Löfven, further advanced that position in October 2015 by making a commitment in a speech to the United Nations General Assembly that Sweden would '*become one of the world's first fossil-free welfare countries*', insisting that '*We are the first generation that can put an end to poverty, and the last that can put an end to climate change*'.[36] This is a bold aspiration, with approximately 50% of Sweden's power still produced from fossil fuels. However, as a statement of moral responsibility for future generations and leadership, it is also a spur to business to innovate climate-smart innovations that will yield benefits environmentally, socially, and economically in an inevitably much changed future world and energy market.

As a converse trend, the USA—a nation that chose not to ratify the Kyoto Protocol—has invested heavily in exploitation of cheap shale gas through the controversial 'fracking' (hydraulic fracturing) process. As of March 2015, the USA had invested $138 billion in new shale gas infrastructure for creating 383,000 direct and indirect jobs by 2023 and £266 billion, new economic output.[37] This contrasts with the position in Europe where a speaker from DG Growth (*Directorate-General:* Internal Market, Industry, Entrepreneurship and SMEs) confirmed that the EU has chosen to take no view of fracking, leaving decisions entirely up to Member States.[38] At the time of writing, France has imposed a total moratorium, perhaps influenced by the profitability of its nuclear generation sector in domestic energy and technology export markets, whilst the UK appears to be placing the immediate economic benefits of fracking over and above any longer-term or systemic implications in the name of a myopic 'growth agenda' blind to its long-term costs. International discrepancies are also causing asymmetries in costs to business, for example, with the US plastics industry benefitting from both cheaper feedstock and energy to produce products that are substantially outcompeting European plastics on international markets. Undoubtedly, uniform standards and regulation are necessary to 'level the playing field' and to factor impacts on ecosystems and future security of ecosystem services into international trade. The mining of shale gas is anyhow an illogical response in the face of international commitments to move away from reliance on fossil carbon-based energy, especially given substantial unknowns about implications for wildlife, aquifers, geological stability, the eventual fate of the large volumes of water and associated additive chemicals used in the fracturing process, and implications for adjacent property values. A further down-side to the release of all this shale gas is that excess US coal is now exported worldwide—about half of it to Europe and a quarter to Asia—resulting in 'dirtier' emissions in these places, further destabilising a global climate that affects the global population now and into the future.[39]

Perhaps too much attention has been dedicated at a global scale to generation, rather than home- and community-scale solutions both to energy generation and also the cleanest energy of all: that which is NOT used due to energy efficiencies. As Feed-in Tariff systems in Germany and, to a lesser extent, the UK have demonstrated, it is possible to drive innovation, price-performance and pervasive uptake of solar and wind generation technologies, and also potentially tidal and water flow-related generation, given the right policy environment. Subsidy schemes too can make economic returns on cleaner technologies such as Air Source and Ground Source Heat Pumps increasingly attractive, as well as energy-averting technologies such as improved insulation and the design of buildings and urban areas for heat retention and/or passive cooling. A broader view of net environmental and societal outcomes from energy generation and use, and progressively closer coherence between those that are more favourable and the broad formal and informal policy environment, could do much to promote more sustainable, ecosystem-friendly energy use practices in both the already developed and the developing worlds, including both driving new innovations and also uptake of technologies that are already available.

The climate system is a global common. It matters not where carbon is emitted, or where it is stored: the environmental, social, and business implications of climate instability affect the whole world, albeit differentially. Politically, innovations such as payment mechanisms from the global north to the global south, reflecting the benefits that industrialised nations have gained from exploiting fossil carbon and the value of retaining reserves in situ and retaining further sequestration potential in the forests, wetlands and soils of developing nations, are beginning to put policy flesh on the bones of high-level agreements. In the light of this, hasty, narrowly self-beneficial exploitation of shale gas is hardly the strongest signal promoting concerted global action for constraining net carbon emissions on an equitable basis. The extent to which energy policy has thus far embraced the ecosystems revolution is therefore at best patchy and generally deeply questionable.

Urban Design

The distinction erected between rural and urban policy across many jurisdictions is, in reality, a false dichotomy. Urban ecosystems are of vital importance to the viability of cities, just as resource flows supporting urban life reach out into rural and increasingly global hinterlands.

Wherever the air that we breathe, the water we drink, and the food we eat comes from, city life absolutely depends on the ecosystems that produce them. Damage or lose those ecosystems, and we lose both these vital ecosystem services as well as others such as natural breakdown of wastes, places outside the city to seek recreation and relaxation, and the natural control of floodwater before it flows into the built environment. Today's ever more globalising world depends increasingly on remote ecosystems.

Cities themselves also comprise a complex mosaic of ecosystems, all of which contribute significantly to urban quality of life, if largely historically overlooked under the false urban-rural dichotomy. For example, the natural topography and permeability of the landscape provides pathways down which storm water flows to regulate urban flooding provided, of course, we respect those natural pathways and do not impede flows or percolation with unsympathetic built infrastructure. Street trees and urban parks give us green spaces that buffer noise and 'hard' cityscape views, break down 'heat islands', and provide places where we can relax, exercise and enjoy nature, individually or with others from our communities. Without nature, or where we fail to make space for ecosystems in city design, we lose these beneficial services provided largely 'for free' by nature that we have, for too long, taken for granted. A cityscape denuded of nature is poorer spiritually, socially, and in terms of human health, but also economically as we suffer elevated risks and/or rely on technological means to compensate for lost or degraded ecosystem services. Heightened flood risk and damage from flooding is a common urban condition and cost where natural hydrology is overlooked; poor public health through bad air quality is another adverse consequence, and limited access to calming green spaces and places to socialise and for healthy exercise reflect additional costs arising from lost services. Air conditioning and its associated energy inputs and operating chemicals help us cope with 'heat islands' in dense cityscapes, as do 'hard' flood defences. These factors not only impose their own 'taxes' on urban centres, but also blight real estate values. Yet great strides have been made with 'green infrastructure' in urban construction—as described in some detail in Chap. 5—including multiuse floodwater infiltration basins, rain gardens and other forms of SuDS, green roofs, and a diversity of novel approaches that can create tangible benefits whilst contributing positively to the 'liveability' and real estate values of the built environment.

A noble attempt at sustainable city design, taking account of multiple dimensions of natural benefit, was in a sustainability strategy developed for a South African 'aerotropolis'—a city extension around a new airport—for which I led the ecosystem services element as part of a multidisciplinary team from the University of the West of England. (The city extension itself had better remain anonymous as our report was to the client only.) Our designs were innovative, for example, advising on urban and industrial developments and supporting infrastructure that followed contour lines around wetlands, not merely meeting statutory requirements but bringing multiple associated ecosystem service benefits into the cityscape including green spaces, natural flood regulation, breakdown of heat islands, corridors for nature and sustainable transport, amenity and recreational areas, and retention of the carbon sequestration, water quality regulation, habitat and natural hazard buffering of the wetlands themselves. Water resources for abstractive and other uses were safeguarded by plans for protecting river quality. Planned natural flow regulation, in addition to traditional wastewater treatment, was also designed to minimise hydrological and chemical changes in the ecologically and recreationally important wetlands and estuary downstream. Our draft design retained trees, and contained provision for planting more as buffers against noise and visual intrusion as well as the settling of fine particulates from the air, and also included tree species providing a source of fuel wood and fruits for poorer members of society integrated into mixed developments. Green corridors permeated the city, bringing with them a host of benefits including corridors for wildlife and sustainable transport. Sadly, production of the strategy seems to have been sufficient for the developer to 'tick the sustainability box', as development proceeded along pretty much traditional lines of square grids conflicting with pre-existing ecosystems and their services. But the example is worth citing as it demonstrates both what can be envisaged on the basis of an intent to embed the values of ecosystem services into design, as well as how bold intentions still conflict with current harsh commercial and planning realities shaped by a short-term profit motive.

Nevertheless, elements of sustainable drainage systems (SuDS)—swales, detention ponds, and basins that may double up as amenity areas, reedbeds, and many more designed to optimise production of multiple services—are becoming more commonplace around the world in urban design. So too, many development schemes are making use of, or are restoring, 'green' and 'blue' (vegetated and waterway) corridors, access to which is associated with many physical and mental health benefits as well as elevated real estate values.

Bold ventures in whole-city sustainable design are also taking shape around the world, including, for example, Masdar City that, at the time of writing, is being constructed 17 kilometres (11 miles) east-south-east of the city of Abu Dhabi near the International Airport.[40, 41] Masdar City is a planned city, relying on energy from solar and other renewable sources and, in time, will host the headquarters of the International Renewable Energy Agency.[42, 43] It is also intended to be a hub for 'clean technology' companies and institutions, the first of which, the Masdar Institute of Science and Technology,[44] has been operating in the city since 2010. Completion of the first phase of the city has been delayed due to the global financial situation, with completion scheduled for between 2020 and 2025. Sustainable innovations include mixed-use development designed to be friendly to pedestrians and cyclists, buildings clustered together with short and narrow streets modelled on other Arab cities to increase shading, terracotta walls, and a 45 metres (148 feet) high wind tower modelled on traditional Arab designs that sucks air from above pushing downwards a cooling breeze that contributes to making street temperatures around 15–20 °C (59–68 °F) cooler than the surrounding desert. Initial plans had excluded automobiles, with travel accomplished using public mass transit and personal rapid transit systems (PRT: essentially small mobile pods). However, PRTs were not extended beyond the October 2010 pilot scheme for cost reasons, with electric vehicles being tested instead for both personal and freight rapid transit purposes. Most private vehicles will be restricted to parking lots along the city's perimeter. Masdar City will connect with Abu Dhabi's existing light rail and metro line, linking it to the greater metropolitan area. Water management has also been planned in as environmentally sound a manner as possible, including extensive SuDS and with 80% of the water used recycled and reused '*as many times as possible*' including for crop irrigation and other purposes.[45, 46] The expected population of up to 4000 in 2014 is predicted to grow over time to 10,000.

In China too, development has proceeded with a range of planned sustainable cities, also referred to as 'ecocities',[47] designed with consideration of environmental impact, and generally inhabited by people aspiring to minimise inputs of energy, water, and food, and the production of waste, heat, and air and water pollutants. Examples of pioneering Chinese ecocities include: Binhai, named the 'Sino-Singapore Tianjin Eco-city' reflecting its principal sponsors[48, 49]; and also Dongtan Eco-city,[50] Shanghai, located in the east of Chongming Island, initially scheduled to accommodate 50,000 residents by 2010 although currently with development on hold.[51]

These and other ecocities planned or under development in various locations across the world have received, predictably, a mixed global reaction. Some commentators see them as tokenistic gestures for the rich, and as 'greenwash' for the states that sponsor them. However, aspirations to achieve zero-waste, zero-carbon urban development have much to commend them, with lessons, hopefully, percolating out not merely to influence new build elsewhere but to the retrofitting of existing urban centres. Today, progress is being made, albeit on a sporadic basis, yet progressive mainstreaming would bring with it demonstration of efficacy and values and, with them, enhanced technical capabilities and acceptance of a more ecosystem-focused basis for urban development.

Under the ecosystems revolution, planning policies should become increasingly aligned to safeguard, and make positive uses of, the services provided by ecosystem assets such as river and stream corridors and drainage lines, seeking multi-benefits including movement of wildlife, cooling and amenity areas, air quality, noise and visual buffers, and routes for sustainable transport methods such as walking and cycling that can avert pollution elsewhere in the cityscape, also providing life-affirming proximity to nature. (The subsequent sub-section addressing *Local government* refers to a research project that explored how an Ecosystem Approach could most effectively and comprehensively be integrated into the urban planning process and its associated tools.)

Policy revision for urban design is also necessary at the home scale. As the German and early UK FIT schemes demonstrate, incentivising a range of off-the-shelf and emerging renewable generation and energy efficiency technologies could trigger step changes in sustainability performance, often cost-neutral at the point of new build. Regrettably, the policy environment today still largely perpetuates the situation in which most of these readily available solutions, more sustainable and cheaper over product life, remain far from mainstream in design and build. This is because the wider policy environment still largely maximises rewards to designers and builders through minimising their capital costs, with risks, operational expenses and lifetime sustainability footprint divested down the value chain to subsequent occupants of inefficient buildings. More could, and should, be done to address these policy disconnections with stated commitments to reduce greenhouse gas emissions and work towards wider sustainable development goals.

Transport

Within and between urban centres and spreading their tendrils out into the rural hinterland is a variety of transport systems. These systems have traditionally failed to make connections with their ecosystem dependencies and, downstream, their broader consequences for the environment beyond growing concerns about climate-active gas emissions. Issues such as the flooding of road and rail infrastructure have instead generally been perceived and managed as a local problem, to be dealt with by installing bigger drainage pipes and/or pumps.[52] Yet there is now a slow waking up to the reality that these transport routes have generally been aligned for convenience, for example, railway lines built as straight as possible across drainage basins and in tunnels that cut through aquifers. The hard lines could have been softened, better to align with the natural processes generated by the topography of the landscape. Instead, road and rail embankments, cuttings and tunnels tend to generate flooding upstream, on the transport infrastructure as well as downstream, whilst also often serving to concentrate pollutants that compromise the quality, utility, and ecology of watercourses. Impacts are substantial, including for the transport system, the host catchment and the ecosystems and people dependent upon it.

Slow progress is being made in redressing some of these impacts, although the 'force majeure' of capital cost often still averts truly systemic thinking about what is best both for the operational efficiency of the transport system and net benefits to society. This is compounded in some territories, such as in the UK, where specific historic Acts of Parliament in effect assign a waiver to operators concerning such 'inconvenient' external considerations. In the field of transport infrastructure, we undoubtedly have a considerable journey to travel under the ecosystems revolution. And this is before we look at the overwhelming dependence on fossil fuels of today's dominant modes of transport, notwithstanding progress with novel mass transport and 'green' multi-use walking and cycle ways in some cities.

Agriculture and Food

Of the many benefits that ecosystems provide for people, food production ranks with provision of clean air and water as amongst the most fundamental. This fact remains true whether we crop food directly from nature, or else modify productive systems as, for example, through farming and aquaculture. All forms of food production depend entirely on a host of natural processes. Reliable flows of fresh water are a vital resource for irrigated and rain-fed agriculture alike, as indeed for stock watering, horticulture and aquaculture, with risks arising if water resources become contaminated or over-exploited. Equally, food production, and increasing pressures on productive land for biofuel, chemical feedstock, pharmaceutical and other novel crops, can displace other important functions and services produced by landscapes and native ecosystems including their medicinal, ornamental, cultural, tourism and other attributes, also compromising the regulation of climate, water flows, air quality and populations of pest, disease, and pollinator species.

Globally, agricultural activities are recognised as amongst the greatest threats to wetland and other terrestrial ecosystems and their broad range of services.[53, 54] A significant part of the problem is myopia about the nature of agriculture best suited to meeting the needs of the disparate global population, blinkered as it is by a dominant capitalist world view framed by profitability and uniformity of methods. Yet it is also the case that innovative as well as some traditional production systems can provide food simultaneously with a range of connected beneficial services as demonstrated by the post-medieval innovation and persistence in places for over four centuries of the water meadow system in Britain[55] (see Chap. 3), or widespread paddy systems across Asia that not only produce a staple crop but retain soils, recycle nutrients and water, bind communities, serve religious functions and support substantial biodiversity.[56]

The 500 million or so smallholder farmers around the world, many of whom are struggling in the face of climate change and economic uncertainty, nevertheless still succeed in feeding one-third of the world's population.[57] This is a remarkable achievement, albeit one that goes largely unrecognised on the international stage, but could be better and more beneficially promoted as a form of 'pervasive localism' as articulated in Chap. 5. With the global population set to exceed 9.5 billion by 2050, increasing the security and productivity of small farms, rather than rolling out a more market-driven model of resource-hungry, genetically uniform agri-business, will doubtless play a vital role in alleviating hunger and poverty and meeting global food demands. As we have seen in Chap. 5, indigenous knowledge and culturally and geographically appropriate local solutions have the capacity to promote the linked regeneration of ecosystems and livelihoods, providing a proven foundation if more widely recognised, supported, and promoted.

The quest for sustainable agriculture, in terms of technologies and common stewardship systems, may be one of the greatest potential contributions to a more sustainable accommodation of direct human needs with the vitality of ecosystems essential for many other purposes, including the long-term viability of agriculture itself.

Health and Wellbeing

There is a close fit between the pursuit of public health and realisation of the Ecosystem Approach. As the World Health Organization (WHO) put it in 2011, '*Health is our most basic human right and one of the most important indicators of sustainable development. We rely on healthy ecosystems to support healthy communities and societies*'.[58] Addressing parameters as diverse as nutrition and food security, clean air and fresh water, medicines, cultural and spiritual values, and contributions to local livelihoods and economic development, the linkages between ecosystem functioning and the wellbeing of people are of unassailable importance. However, current fragmentation of the policy environment frustrates the mainstreaming of beneficial ecosystem services across policy areas as a net contribution to human wellbeing and sustainability. The mapping of the services provided by ecosystems to multiple human wellbeing end points by the Millennium Ecosystem Assessment[59] is widely accepted, highlighting how access to adequate supplies of natural resources and sound governance of them are key arbiters of harmonious and potentially sustainable, or alternatively competitive and conflicted, human interactions at a range of scales from the local to the international. Convergence on the term 'wellbeing' in the languages of health and ecosystem services recognises their respective multidimensionality, both in terms of the need to recognise a plurality of human needs but also of factors contributing to them. Thus, the ecosystem services model renders more tractable the articulation of why and how ecosystems and their vitality are important to people, supporting overall public health and wellbeing.[60]

An important conceptual challenge remains in terms of integrating 'health' (in its broadest sense) into the Ecosystem Approach, including determining with far greater resolution how health end points map against the established Millennium Ecosystem Assessment framework of provisioning, regulatory, cultural, and supporting ecosystem services. It is likely that 'health' itself is a meta-outcome of many interlinked services. The UK's most widely accepted conceptual model of public health is the *Public Health Outcomes Framework 2013 to 2016* published by the Department of Health.[61] This public health model recognises two principal outcomes: increased healthy life expectancy; and reduced differences in life expectancy and healthy life expectancy between communities. These sit within an overall vision, supported by four 'domains': improving the wider determinants of health; health improvement; health protection; and healthcare, public health and preventing premature mortality. The overall approach focuses not merely on disease prevention but upon the rebuilding of human resilience as a basis for health generation, characterising an emerging agenda of 'salutogenesis', a term coined in 1979 to describe an approach focusing on factors that support human health and wellbeing, rather than on factors that cause disease.[62] There are clearly multiple direct linkages between ecosystem services and these four public health domains, ranging from the more obvious example of air quality regulation to less direct, but nevertheless clearly demonstrable health outcomes related to access to green spaces and healthy exercise. Deeper connections between ecosystems and human health are described by evolutionary biologist E.O. Wilson, who coined the term 'biophilia' to describe necessary connections between humanity and the environment in which we evolved as a basis for sound physical and mental health.[63, 64] In 2005, the WHO produced a 'Health synthesis' as part of the Millennium Ecosystem Assessment,[65] drawing a number of conclusions highlighting the direct linkages between ecosystems and health, recognising that '*Ecosystem services are indispensable to the well-being and health of people everywhere*'.

Public health may not, to date, have been an organising principle for all facets of the operation of civil society. However, every management decision or operational intervention across societal policy areas—such as energy and transport systems, built environment design, food production methods and the food itself—has potentially positive or negative consequences for all ecosystem services and their associated beneficiaries. There is therefore an increasing opportunity to integrate public health outcomes into all forms of ecosystem management and public policy; conversely, overlooking public health implications can potentially degrade them in pursuit of the maximisation of more narrowly framed service outputs.

Culture

Many of the points addressed above with respect to the links between ecosystems and health are relevant to the pervasive and profound contribution of the natural world to culture. Water, soil, microclimate and distinctive landscapes and species of trees and other organisms, for example, all contribute substantially to local distinctiveness and culturally valued locations. From these may also stem substantial spiritual, artistic, healing, and tourism benefits. These benefits, of course, rest upon a wide range of supporting and regulatory services, often overlooked but always contributing to the natural heritage that enriches and connects society.

Cultural benefits from nature have fared mixed fortunes in terms of the extent to which they have been captured by markets and other policy responses. Exploitation of some habitats for recreation and tourism yields growing value on the basis of clear monetary implications, although also potentially placing pressure upon them, and some sites of spiritual and/ or heritage value receive explicit protection. However, landscapes change constantly in response to shifting environmental conditions, land uses, and societal priorities. The range of meanings that landscapes confer upon different stakeholder groups is often poorly represented in decision-making. A gross example of this is the inundation of culturally meaningful sites through the filling of large dams, multiple displacements of communities who do not share the benefits of dam development, and the often sizeable constituencies whose livelihoods are adversely affected by the changing character and services of the host river system.[66, 67, 68] Degradation of any type of ecosystem can systematically undermine the physical health and socio-economic wellbeing of communities, including their long-term viability. This chain of cause-and-effect is not often recognised in cultural protection.

Significant within a broader consideration of cultural reform is the role of the world's religions. A comprehensive review of the role of *Faith in Conservation*[69] reveals, amongst the highly heterogeneous belief systems and traditions of 11 of the world's major religions, common threads of responsibility for the stewardship of nature in its widest sense, including not only its inherent values but also as a shared resource for all of humanity. In Catholicism, the 2015 Papal Encyclical *Laudato Si*[70] explicitly acknowledges this planet as '*our common home*' that now '*cries out to us because of the harm we have inflicted on her by our irresponsible use and abuse*', calling on the world's rich nations to begin paying their '*grave social debt*' to the poor and particularly to take concrete steps on climate change. Opposition to the pope's statement from right wing politicians in America only endorses the insightfulness of the Encyclical in highlighting how advantaged sectors of the global population continue to exert the greatest pressure on already stressed common planetary ecosystems and thereby upon those people most directly affected by climatic and other ecosystem changes. Certainly, in promoting the fulfilment of human potential, the vision, very substantial resources, political and domestic influence and educational activities of the world's religions need further to embrace contemporary biospheric challenges and their positive solutions, representing potentially significant agents driving ahead the ecosystems revolution.

Local Government

Given its influence on urban planning, local economic development and public health, local government is highly significant in framing relationships between ecosystems and human activities occurring within them. Local government decisions are significant in shaping how urban and industrial development interacts with a range of ecosystem processes—hydrological, chemical including the cycling of nutrients and carbon, regeneration of soils and living organisms, purification of air and water, and so forth—vital to the sustainability of the intimately interconnected socio-ecological system.

Owing largely to a shortfall in awareness of these intimate interdependences, local government decision-making has resulted in a long-running legacy of planned and permitted development favouring excessive or inappropriate schemes. This in turn has tended to undermine health and wellbeing, the 'liveability' of the built environment and to contribute to climate instability and other ecosystem functions compromising long-term viability. However, given this range of influences, local government decision-making is also potentially significant in influencing a more sustainable pathway of development, notwithstanding a substantial body of vested interests and flawed assumptions that need progressively to be challenged and changed. Both the optimisation of what is possible as well as the lamentable reality of current vision and lax enforcement are exemplified by the South African aerotropolis example outlined above under *Urban design*.

However, ideally supported by a favourable policy environment shaped by central government, local government can play a significant role in encouraging, through presumptions in planning policies and other easements, of the locally beneficial uptake of the kinds of environmentally and societally beneficial home- and community-scale energy generation and efficiency technologies discussed under consideration of *Energy* above, as indeed water- and wildlife-friendly plans and policies.

Embedding an Ecosystem Approach as the basis for local government decision-making offers the long-term promise of safeguarding and restoring or emulating natural heritage and processes into development schemes. This form of directed, multifactorial approach best safeguards public health and enriches local communities through both more tangible aspects, such as elevation of real estate values and reduction in flood risk, and less tangible but nonetheless important ways such as enhancing community identity and 'liveability'. A major strand of the UK's National Ecosystem Assessment follow-on programme addressed how best to embed the Ecosystem Approach into the planning cycle, and a wide range of decision-support tools used within it.[71] The embedding of the Ecosystem Approach into the planning process is very much a 'work in progress'. However, it is one that has significant promise in driving ahead a systemically interconnected ecosystems revolution spanning and influencing all policy areas, for the benefit of those the system is intended to protect and support.

Natural Environment

And so we arrive at consideration of the natural environment in policy and business deliberations. The common phrase 'environmental management' is, in many ways, a curious and outmoded oxymoron. After all, the environment—literally that which surrounds us—has always adapted to changing cosmological, anthropogenic, and other pressures, and will always do so in whatever form. It is people, and what we do, that need managing if we are to safeguard a future that, on current trends, looks to be increasingly impoverished by degradation of the ecosystems that support but also limit its potential. The strategic challenge is to embed the workings and benefits of the natural environment into all other policy areas, rather than to persist with 'environmental management' retrospective to narrowly framed economic decision-making, as if policy and practice across society operated in some virtual hyperspace divorced from dependencies and impacts on the natural world. As famously and elegantly framed by Dr Herman E Daly, widely acknowledged as the father of the field of ecological economics, '*The economy is a wholly owned subsidiary of the environment, not the reverse*'. Yet all too often, the 'environment department' within policy, business, municipality, and other institutions is only brought into play to deal with the aftermath of impacts arising when primary decisions have already been made, and/or have so much investment of financial and/or political capital that they are largely resistant to significant amendment. Automatically, ensuing responses are geared to damage limitation rather than safeguarding vital resources. And, because of this externalisation of the natural world and its multiple beneficial services from primary decisions, nature conservation obligations and other 'environmental' measures are often seen as inconvenient constraints on legitimate social and economic development rather than as a fundamental investment in primary resources assuring their sustainability.

Biodiversity, both within the natural world and as agrobiodiversity, has a fundamental role to play in sustainable transformation, constituting the foundations of many ecosystem services and the basis of the Ecosystem Approach. Consequently, environmentalists too have to challenge their own established assumptions and practices, expanding thinking beyond historic preservationist approaches to nature conservation into something a lot more systemic. Specialised protection of the most threatened species and habitats remains a priority. However, historic approaches to 'fortress conservation' can prove unhelpful in promoting the value of nature if conservation is perceived as lacking in social and economic context and consequently, as is common today, regarded as an inconvenient constraint on legitimate economic progress. Conservationists need to remain strident in championing the inherent value of nature, but also to recognise and promote the beneficial processes to which organisms and habitats contribute, thereby championing the systemic socio-economic values of nature.

The agency of 'nature', in its broadest sense, in sustaining and regenerating human opportunity and endeavour in rural and urban, developed and developing worlds alike, is exemplified in many examples of *Reanimating the landscape*.

A central plank of the necessary culture change is recognition of 'the environment' not as something to be considered once the big commercial and other decisions have been made, but as comprising fundamental resources valued for the multiple benefits they provide as an integrated part of primary investment decisions. Indeed, the priority given to the supportive capacities of the natural world in the decision-making cycle is a key indicator of societal progress with the ecosystems revolution.

International Development

Unlike domestic policy, international development initiatives have for a matter of decades largely embraced aspects of an Ecosystem Approach. This applies in particular to development support projects in regions with a predominance of resource-dependent economies, a definition used for those most directly reliant on primary natural resources to supply their immediate basic needs (albeit a phrase that betrays our industrialised world naivety about the fact that we are ALL resource-dependent). Progressive international development support initiatives have long recognised that the rebuilding of natural capacity, the modification of resource use practices consistent with natural limits, and the restoration of often traditional local governance arrangements lie at the heart of resource security upon which durable social and economic progress is founded. Classic examples include the linked socio-environmental regeneration programmes in semi-arid north Rajasthan, China's Loess Plateau, and Highland Ethiopia, reviewed in Chap. 5.

A linked set of water, soil health, and food productivity concerns has often framed such types of development initiatives, contributing to livelihood security and generally also seeking additional economic viability. Typically also, this has been in response to serial degradation of natural capacities through over-exploitation or uses unsympathetic with the character of the ecosystems with which people live, precipitating drought, famine and/or distress migration or lost opportunity. With the loss of primary resources has come degradation of natural regulatory services exacerbating soil erosion, extremes of flooding and drought, local and wider climate variability and vulnerability, and problems with pests, diseases and natural processes such as nutrient retention and pollination compromising eco-system integrity and crop productivity. Aid is best delivered by helping people address their own problems through sustainable resource steward-ship, often supporting them in instituting traditional or novel stewardship and local governance regimes; the historic approach of merely buying in commodities has been largely abandoned today outside of relief from the acute aftermath of disasters.

Since 2010, there has been increasing international interest and activity around a broader 'nexus' of issues, most commonly articulated as closely interlinked water, food and energy issues, including addressing the chal-lenges of a changing climate. This is often considered in the context of emancipation of the 'bottom billion' of the world's population from mul-tiple, interlinked dimensions of poverty, as expressed in the Millennium Development Goals[72] and their successor Sustainable Development Goals.[73] However, this nexus of issues and the need for a systematic response to address it is also a pressing challenge for securing resilience of the already developed world.[74] Indeed, the nexus concept is beginning to influence the thinking of some leading industrial sectors recognising that water, food, and energy are pieces of the same puzzle, and that it is there-fore not practical to look at them in isolation.[75]

Often, as in the case of the Ecosystem Approach itself, the paradigm-changing ideas that are progressively pervading all sectors of human inter-est, and that need to be accelerated to safeguard the interests of all in an intimately integrated global society, stem from insights and innovations arising from places where ecosystems and the wellbeing of people are most directly tied to each other.[76]

Defence

Conflict is always about something which, even when dressed as a matter of ideology, race, or belief, usually has competition for a limited resource at its core. Oil has been a commonly contested resource in our more recent industrial past. The 2009 prediction by then World Bank Vice President Ismail Serageldin that '*Many of the wars of the 20th century were about oil, but wars of the 21st century will be over water unless we change the way we manage water*'[77] is today still often repeated, yet also betrays naivety about the long historic legacy of water-related conflicts. As one example from a half-century ago, the Six Days War of 1967 had competition for the waters of the Jordan River at its heart. The flashpoint was commencement by the Arab League of the digging of canals to divert two of the major tributaries of the Jordan River, the Hasbanin, and Wazzani Springs, in an angered response to Israel's construction of its National Water Carrier project that appropriated much of the flows of the Jordan River. The two Arab League schemes were summarily shelled and destroyed by Israeli forces, precipitating retaliatory attacks by Syria, Egypt and Jordan. In the ensuing war, Israel seized Gaza and the Sinai peninsula from Egypt, the Golan Heights from Syria and the West bank from Jordan, all but the Sinai peninsula important as catchment sources supporting Israel's water security. This 'water war' is just one of many examples that include drainage of the Mesopotamian wetlands as a potent weapon of war waged by Saddam Hussein's forces against Iraq's Marsh Arabs, and indeed behind simmering concerns about the 'unquenchable thirst' of India, Pakistan, and China over the waters and power generation potential of major transboundary rivers fomenting rivalries threatening South Asia's peace.[78]

However, if water and other natural resources are a potential spark for conflict, they are also a catalyst for peace-making and additive cooperation. The 2006 UN report *Water Wars to Bridges of Cooperation—Exploring the Peace-Building Potential of a Shared Resource*[79] provides a valuable catalogue of the role of water resource-sharing and co-management as a prevalent contributor to international security across the globe. Many more examples of water-based cooperation with associated peace-keeping and peace-making are reviewed in my 2013 book *The Hydropolitics of Dams.*[80]

The key point in considering defence strategies as part of the ecosystems revolution is that water and other vital natural resources cannot be dissociated from the securitisation agenda, including avoidance of conflict and in peace-making and the securing of peace after conflict. This reality is not lost on the military. For example, the UK Ministry of Defence's Development Concepts and Doctrine Centre (DCDC) is tasked with forming a longer-term view of global trends with security implications, and it takes a strong interest in resource security as a means to avert conflict and underwrite peace post-conflict. Indeed, resource security features prominently and frequently throughout DCDC's *Global Strategic Trends—Out to 2040*[81] review. Increasingly, ecosystem assets and their contribution to securing the 'nexus' will feature in how conflict is averted and overcome, and how peace is secured on a sustainable basis thereafter.

Foreign Policy

Aspects of foreign policy have been addressed above in terms of trade and industry, energy, international development, and security. All could become better informed by recognition of dependencies of local people and supply chains on ecosystems, and equally about the consequences of this for ecosystem resilience. However, another important aspect of foreign policy is in the way that the international community can mobilise around global threats.

My 2015 book *Breathing Space*[82] addresses in detail how the global community has acted to respond to a range of atmospheric pollution issues. Significant amongst these was the response to growing international concerns about elevation of radioactive contamination of the atmosphere, culminating in the Limited Test Ban Treaty in 1963 which also served to slow down the dangerous pace of the nuclear arms race and open the way to disarmament. Concerns about transboundary acid rain issues drove further international action in the 1970s between nations across Europe, also addressing reciprocal damage between Canada and the USA, leading to development of the UNECE (United Nations Economic Commission for Europe) Convention on Long Range Transboundary Air Pollution adopted in 1979 and entering force in 1983.[83] Discovery of Antarctic ozone thinning in 1985 was another stimulus for concerted global action to fix this 'hole in the roof of the world', unilateral action by a range of developed economies culminating in 1987 in the *Montreal Protocol on Substances that Deplete the Ozone Layer*.[84] Although this multinational agreement is far from perfect, Kofi Annan, former Secretary General of the United Nations, stated that '*perhaps the single most successful international agreement to date has been the Montreal Protocol*'.[85] Further international responses include numerous treaties on the co-management or allocation of the flows of transboundary rivers. Various European Union Directives also impose common environmental standards and protective measures across European Member States. The 1971 Ramsar Convention on *Wetlands of International Importance*[86] remains the only global convention dedicated to a specific habitat type, and has served to some extent to safeguard important networks of wetlands across continents but also to champion the retention and restoration of important wetland ecosystem functions in farmland and urban development as a powerful means to embed an Ecosystem Approach into these areas of human progress.

Perhaps the most significant global mobilisation around a looming hazard is seen in response to the planet-wide and long-term threat of climate change. Prime amongst international cooperation platforms is the Intergovernmental Panel on Climate Change (IPCC),[87] set up in 1988 at the request of member governments as a scientific intergovernmental body to provide comprehensive assessments of current scientific, technical and socio-economic information worldwide about the risks of climate change caused by human activities, their potential environmental and socio-economic consequences, and possible options for adapting to these consequences or mitigating their effects.[88] IPCC knowledge in turn supports deliberation and agreements under the United Nations Framework Convention on Climate Change (UNFCCC[89]), established in 1992 as an international forum for negotiation on adaption to climate change, financing of action on mitigation and adaptation, mitigation of greenhouse gas emissions, and technology development and transfer to allow green development.

International cooperation and obligations have been powerful drivers of change, the Ecosystem Approach itself a product and agreement under the intergovernmental Convention on Biological Diversity,[90] even if the slow pace of the ensuing ecosystem revolution remains a source of frustration and net cost to both current and future generations.

Research and Education

Sustainable and unsustainable activities alike across societal interests offer valuable learning and research opportunities, if we have the capacity to discern them. And this cuts back to what we learn, and what we fund in terms of research.

On these scores, we suffer still from too great a fragmentation into disciplinary specialisms both in research and other branches of education. Of course, specialist focus is necessary in a complex world wherein we need those who can think through difficult ecological, chemical, engineering, planning, fiscal, and other problems. However, what remains distinctly lacking is the 'glue' that joins these different sectoral interests together, for all are distinct only as a matter of anthropocentric definition; all constitute different facets of the same interconnected natural, human, and economic world. For example, food production systems have wide and largely unintended ramifications for water quality and quantity in catchments, the vitality and microbiology of soils, nutrient cycles, biodiversity and recreational potential, and hence the longer-term viability of food production, the economic performance of the food sector including its contribution to international competitiveness, and hence the regulation, taxation and subsidies that best sustain the cyclic whole. Treat partitions in this continuum reductively, as we so commonly do today, and the cycle breaks down both biophysically and economically.

Consequently, the ecosystems revolution demands that systems thinking becomes integral to the true understanding and contextualisation of specialised knowledge, for without knowing how our jigsaw pieces fit together and influence each other we risk remaining 'bit part players' potentially eroding the greater good through too narrowly framed best intentions.

As big a challenge is presented by the growing psychological and behavioural disconnection of people, younger people in particular, from the ecosystems of which they depend. This leads to what is increasingly recognised as 'nature deficit disorder', a term coined in 2005 by Richard Louv,[91] descriptive of the human costs of alienation from the natural world due to spending less time outdoors. Whilst 'nature deficit disorder' is not a medical condition, it can be contributory to health issues—such as short-sightedness from excessive time looking at computer screens rather than at distance, or increased obesity and diabetes resulting from a sedentary life style—as well as decreasing awareness of the importance of natural processes and services to contemporary lives. In 2012, the National Trust published a research report, *Natural Childhood*,[92] presenting compelling evidence that the UK as a nation, especially children, are exhibiting symptoms of 'nature deficit disorder' due to increasingly sedentary, indoor lifestyles, and proposing activities to reconnect younger people with the natural world. As already addressed above when considering *Health and wellbeing*, there are wider ramifications of this broadening disconnection from the natural world and the processes that sustain us. Perhaps our ultimately destructive contemporary economic model, seemingly hell-bent on consuming the very natural foundations that have propelled and are essential to sustain it and its potential to enhance human security and equity into the future, is the ultimate monument to humanity's current divorce from nature. Resetting these foundations, from early years education but also across all age groups and societal sectors, is surely a priority for education and research if sustainable foundations are to be laid on the necessary voyage of the ecosystems revolution.

Systemic Connections Across Policy Areas

The above reappraisal of research and education frames much of what is necessary to integrate societal activities such that their systemic outcomes form the basis for decisions right across policy areas, with disciplinary distinctions merely an anthropocentric convenience. All management decisions and practices influence ecosystems in some way, whether directly or indirectly, and so have inevitable systemic consequences for all services and their beneficiaries. It is therefore necessary to be aware, and to then take account, of these systemic ramifications if decisions are to result in consequences that avert negative externalities and are consequently equitable across stakeholder groups, sympathetic with the continued functioning and resilience of supporting ecosystems, and thereby deliver optimal net societal value.

Against this ideal of a truly integrated approach, it is clear that much work needs to be done to break policy formulation out of narrow disciplinary thinking that is generally blind to systemic ramifications. The reality today is that, at higher policy levels in government departments, narrow disciplinary considerations backed up by a legacy of equally narrow regulations, taxes and subsidies and the drag effect of their associated vested interests and assumptions still tend to remain blind to dependencies and impacts on ecosystems. We are also still continuing to export this nature-blind, developed world model of 'progress' with the advance of capitalism and the promotion of unsustainable lifestyle aspirations to the globally booming, resource-hungry middle class of emerging economies. Surely we should aspire to create a better example of sustainable progress, initially as a model to realise ourselves and thereafter to promulgate as a more durable and desirable aspiration to the developing world?

Wholesale change is undoubtedly necessary if the collective lifestyles of humanity are not to continue to unravel the supportive capacities of an already much degraded global ecosystem, which can result only in lost opportunity and ensuing human misery. The task then is to generate a clear future vision from which to direct and select the evolutionary steps that may contribute cumulatively to an enduring ecosystems revolution of expanding capacity and opportunity.

THE BROADER POLICY ENVIRONMENT

It was not my intent in the preceding sections of this chapter to infer that policy stems only 'top-down' from government and its departments. The clustering of policy areas generally reflecting the divisions in most governments was merely a device to stratify different dimensions of human interest. In fact, as we find in the review of the 'quiet revolution' to date, leadership comes from many sectors of society. NGOs may focus on emergent concerns to elevate them to public scrutiny and ultimately response by business and governments, businesses may innovate more ecosystem-sensitive solutions and then petition governments to regulate, subsidise or tax to create a business environment that deselects unsustainable options, common law is a powerful mechanism to protect private and public interests in the light of evolving knowledge, and these 'societal levers' may culminate in new regulatory and/or fiscal instruments to formalise shifts in perception and value into societal norms. These diverse societal levers are in practice highly interdependent and can freely morph into each other, progressively contributing to the transformation of societal norms with regard to different aspects of our relationships both with each other and with the supporting environment.[93, 94]

A strand of research under the UK National Ecosystem Assessment Follow-on programme took an overview of the diverse range of 'response options' available for decision-makers to manage environmental change and meet sustainability objectives including, *inter alia*: top-down statutory regulation and levies; bottom-up initiatives including quality assurance networks or community-based partnerships; formal incentives; and voluntary market-based schemes such as 'payments for ecosystem services' (PES) or offsetting.[95] The research explored the distinct set of characteristics of each type of response option, highlighting their strengths and weaknesses particularly regarding adaptation to long-term change and handling of uncertainty, and their differential suitability to different contexts. From this classification, a typology was developed to provide a reference for recognising complementary rather than conflicting interventions as guided by the holistic principles of the Ecosystem Approach.[96]

Given the diversity of 'societal levers' (or 'response options'), their interactivity and the many vectors through which change can be promulgated, it is dangerous to expect leadership to come solely from 'on high'. As we have seen, legislative responses are often retrospective, frequently stemming from a desire for regulatory controls from other sectors of society, and they are often also developed only slowly and generally with significant trade-offs between competing interests. If we are then to accelerate change towards a more systemic and sustainable relationship with ecosystems vital for continuing human wellbeing and opportunity, it is essential to reach out beyond politicians and the wider media, with greater awareness, innovation and activism across all sectors of society as all have unique perspectives and complementary roles to play in the necessary societal transformation. It is through community-spanning networks—not merely professionals, politicians and civil servants with environmental remits, but all across central and local government, those innovating in business, consultants, academics, and students, NGOs and opinion-formers in civil society—that evolutionary innovations in consciousness, products and world views reverberate and crystallise into new cultural norms.

Systemic and sustainable change necessarily includes mainstream internalisation across policy areas of the economics of nature. Valuation of natural assets and services has its strong critics. A common argument in that valuation attempts to solve sustainability problems by playing the market system at its own game, inevitably falling foul of the desire of a world driven to make profit and the accumulation of wealth as an end-game in itself as capitalism constitutes a barrier to sustainability.[97] Conversely, others argue that capitalism is a pervasive ideology that we not only have to work with, but that can work for sustainable development if the values of nature are progressively integrated into it as the most basic and foundational, yet currently massively externalised, form of capital.[98, 99] It is frequently my experience that critics of taking an economic approach to ecosystems and their services make a combination of any of three flawed assumptions. The first flawed assumption is that an economic approach to ecosystems and their services is about putting a 'price on nature' to be traded in the economy; the reality is that it is not about valuing nature itself, but instead about recognising the multiple values that ecosystems provide as an important input to better informed decision-making. The second common false assumption arises from confusion between valuation and accountancy. Whilst accountancy is narrowly framed by cash values and arithmetic judgements, a proper understanding of valuation is, as reflected by the qualitative differences in ecosystem service categories defined by the Millennium Ecosystem Assessment,[100] necessarily pluralistic in approach, recognising often incommensurable value systems. Although these different types of value are often subsequently normalised in decision-making frameworks, commonly into a monetary value, this is primarily to indicate their relative magnitudes, better enabling them to be recognised and weighed in inevitably political decisions. The third flawed assumption is that we do not already value nature; in fact, we routinely value ecosystems and their services in decision-making today, by and large using a default value of zero, which is something we should and must improve upon in however imperfect a way.

We need a vision of a different society in which nature is integrated into decision-making including, for example, in transport, infrastructure, energy, employment, education, and all other societal spheres. How this can be achieved without some form of valuation as a basis for disbursing investment, shaping decisions, and explaining the basis for them is hard imagine, as the evidence to date is that ill-defined 'inherent value' has little traction in societal decision-making. Furthermore, relying on protection in nature reserves alone wherein nature has a theoretical infinite value, be those reserves bounded physically or in the ways we largely inadvertently mentally compartmentalise nature from the wider economy, serves only to fragment what little is left and to legitimise its demise beyond the boundary fence.

Today, we are a long way from this idealist plural and integrated approach to valuing nature and people in the workings of the market and other levers across society's broad formal and informal policy environment. Progressive recognition and incorporation of the multiple values of natural systems into all facets of societal decision-making is a crucial component of the necessary transition, constituting another defining feature of the ecosystems revolution.

NOTES

1. Everard, M. and Appleby, T. (2009). Safeguarding the societal value of land. *Environmental Law and Management*, 21, pp. 16–23.
2. Everard, M. (2011). *Common Ground: The Sharing of Land and Landscapes for Sustainability*. Zed Books, London.
3. Everard, M., Dick, J., Kendall, H., Smith, R.I., Slee, R.W., Couldrick, L., Scott, M. and McDonald, C. (2014). Improving coherence of ecosystem service provision between scales. *Ecosystem Services*, DOI: 10.1016/j.ecoser.2014.04.006.
4. Huber, J. and Robertson, J. (2000). *Creating New Money*. New Economics Foundation.
5. Galbraith, J.K. (1982). Recession Economics. *New York Review of Books*, 29(1), 4 February 1982.
6. Daly, H. (1996). *Beyond Growth*. Beacon Press.
7. Carley, M. and Spapens, P. (1997). *Sharing the World: Sustainable Living and Global Equity in the 21st Century*. Routledge.

8. Bruges, J. (2000). *The Little Earth Book*. Alastair Sawday Publishing C. Ltd.
9. Nordhaus, W.D. and Tobin, J. (1972). *Is Growth Obsolete?* Economic Growth, National Bureau of Economic Research, no 96, New York.
10. Daly, H. and Cobb, J. (1989). *For the Common Good*. Beacon Press, Boston.
11. Bensel, T. and Turk, J. (2011). *Contemporary Environmental Issues*. Bridgepoint Education.
12. Prahlad Shekhawat, P. (undated). Redefining Progress: A report from the Gross National Happiness conference in Bhutan. Policy Innovations. (http://www.policyinnovations.org/ideas/briefings/data/000098/:pf_printable, accessed 5 October 2015).
13. Talberth, J. (2006). *Sustainable Development and the Genuine Progress Indicator*. Contribution to beyond GDP Virtual Indicator Expo. (http://ec.europa.eu/environment/beyond_gdp/download/factsheets/bgdp-ve-gpi.pdf, accessed 5 October 2015).
14. Helliwell, J., Layard, R. and Sachs, J. (2013). *World Happiness Report 2013*. United Nations Sustainable Development Solutions Network, 9 September 2013. (http://unsdsn.org/wp-content/uploads/2014/02/WorldHappinessReport2013_online.pdf, accessed 5 October 2015).
15. OECD. (Undated). *Better Life Index*. (http://www.oecdbetterlifeindex.org/#/11111111111, accessed 5 October 2015).
16. Eisenstein, S. (2011). *Sacred Economics: Money, Gift, and Society in the Age of Transition*. Evolver Editions, New York.
17. Eisenstein, C. (2007). *The Ascent of Humanity*. Panenthea Productions, Harrisburg, PA.
18. The Co-operative Bank. (undated). *Ethical Policy*. (http://www.co-operativebank.co.uk/aboutus/ourbusiness/ethicalpolicy, accessed 16 September 2015).
19. Triodos Bank. (undated). Triodos Bank. (www.triodos.co.uk, accessed 16 September 2015).
20. MFI Benchmark Analysis. *Microbanking Bulletin*, December 2009 (IssueNo.19).(http://www.themix.org/publications/microbanking-bulletin/2009/12/mfi-benchmark-analysis-microbanking-bulletin-december-200, accessed 16 September 2015).
21. Obama, B.H. (2008). *The Audacity of Hope: Thoughts on Reclaiming the American Dream*. Canongate Books.

22. Bill and Melinda Gates Foundation. (undated). Bill and Melinda Gates Foundation. (http://www.gatesfoundation.org, accessed 29 September 2015).
23. Porritt J. (2005). *Capitalism as if the World Matters.* Earthscan: London.
24. Everard, M. (2000). Aquatic ecology, economy, and society: the place of aquatic ecology in the sustainability agenda. *Freshwater Forum,* 13, pp. 31–46.
25. Eves, C., Couldrick, L., Everard, M., Reed, M., Carlisle, D., McNab, D., White, A. and Smith, S. (2014). *Payments for Ecosystem Services: Developing the Evidence Base on PES Beneficiaries in England—Draft Final Report.* Report by URS, Defra R&D programme.
26. Everard, M. (2009). *The Business of Biodiversity.* WIT Publishing, Hampshire. 284 pp.
27. Forest Stewardship Council. (Undated). *Forest Stewardship Council.* (www.fsc-uk.org, accessed 4 August 2015).
28. Marine Stewardship Council. (Undated). *Marine Stewardship Council.* (www.msc.org, accessed 4 August 2015).
29. Rainforest Alliance. (Undated). *Rainforest Alliance.* (www.rainforest-alliance.org, accessed 4 August 2015).
30. Soil Association. (Undated). *Soil Association.* (www.soilassociation.org, accessed 4 August 2015).
31. Fairtrade International. (Undated). *Fairtrade International.* (www.fairtrade.net, accessed 4 August 2015).
32. Everard, M. (2015). *Breathing Space: The Natural and unnatural History of Air.* Zed Books, London.
33. Energiewende. (Undated). *Energy Transition: The German Energiewende.* (www.energytransition.de, accessed 26 August 2015).
34. Friedman, T.L. (2015). Germany, the green superpower. *International New York Times,* Thursday 7 May 2015.
35. Ofgem. (2011). Feed-in Tariff (FIT) Update Newsletter Issue 4. (http://www.ofgem.gov.uk/Pages/MoreInformation.aspx?docid=13&refer=Sustainability/Environment/fits/Newsletter, accessed 14 April 2015).
36. Wang, A.X. (2015). Sweden, an industrialised country of 10 million, is going all out to ditch fossil fuels. (http://qz.com/520974/sweden-an-industrialized-country-of-10-million-is-going-all-out-to-ditch-fossil-fuels/, accessed 15 October 2015).

37. American Chemistry Council. (2015). (http://www.american-chemistry.com/Policy/Energy/Shale-Gas/Shale-Investment-Infographic.pdf, accessed 11 May 2015).
38. Cozigou, G. (2015). Presentation: *The circular economy, climate and European industry*. VinylPlus Vinyl Sustainabilty Forum 2015, Cannes.
39. Razzouk, A.W. (2013). Fracking: American dream, Chinese pipe-dream, global nightmare. *The Independent*, 13 August 2013. (http://www.independent.co.uk/voices/comment/fracking-american-dream-chinese-pipedream-global-nightmare-8759128.html, accessed 28 December 2013).
40. The Economist. (2008). Masdar plan. *The Economist*, 4 December 2008.
41. Walsh, B. (2011). Masdar City: The World's Greenest City? *Time*, 25 January 2011.
42. Al Lawati, A. (2009). UAE to host Irena HQ. *Gulfnews.com*, 29 June 2009.
43. Lawton, M. (2009). Renewable energy agency to call United Arab Emirates home. *Deutsche Welle*, 29 June 2009.
44. Madsar Institute. (Undated). *Madsar Institute.* (http://www.masdar.ac.ae/, accessed 4 August 2015).
45. Dilworth, D. (2007). Zero Carbon; Zero Waste in Abu Dhabi. *BusinessWeek*, August 2007.
46. Palca, J. (2008). Abu Dhabi Aims to Build First Carbon-Neutral City. *National Public Radio*, 6 May 2008.
47. Register, R. (1987). *Ecocity Berkeley: Building Cities for a Healthy Future.* North Atlantic Books, Berkeley.
48. Singapore Government. (Undated). Sino-Singapore Tianjin Eco-city: A Model for Sustianable Development. (http://www.tianjinecocity.gov.sg, accessed 16 August 2015).
49. UNEP. (2013). UNEP South-South Cooperation Case Study: The Sino-Singapore Tianjin Eco-City: A Practical Model for Sustainable Development, March 2013. (http://www.unep.org/chinese/south-south-cooperation/case/casefiles.aspx?csno=114, accessed 4 September 2015).
50. McGray. M (2008). Pop-Up Cities: China Builds a Bright Green Metropolis. *Wired Magazine*, 15 May 2008.
51. Malcolm M. (2008). China's Dongtan demise is mirrored by bad news for Britain's eco-towns. *The Telegraph*, 18 October 2008.

52. Quinn, N.W., McEwen, L.J., Parkhurst, G., Parkin, J., Everard, M., Horswell, M., McInnes, R.J., Newman, R. and Williams, D. (2015). *Co-creating Railway Flood Resilience: Applying the science of blue-green-grey infrastructure.* NERC 'Environmental Risks to Infrastructure Innovation' programme, University of the West of England.
53. Millennium Ecosystem Assessment. (2005). *Ecosystems & Human Well-being: Water and Wetlands Synthesis.* World Resources Institute: Washington DC.
54. Millennium Ecosystem Assessment. (2005). *Ecosystems & Human Well-being: Synthesis.* Island Press: Washington DC.
55. Everard, M. (editor). (2005). *Water meadows: Living treasures in the English landscape.* Forrest Text, Ceredigion. 289 pp.
56. Pearce, F. (2004). *Keepers of the Spring: Reclaiming Our Water in an Age of Globalization.* Island Press, Washington DC.
57. Birch, S. (2012). Springing the poverty trap. *Green Futures Special Edition: Shared Future—how co-operatives can reboot a sustainable economy,* February 2012, pp. 8–11.
58. WHO. (2011). *Discussion Paper: Our Planet, Our Health, Our Future—Human health and the Rio Conventions: biological diversity, climate change and desertification.* World Health Organization, Public Health & Environment Department (PHE), Geneva.
59. Millennium Ecosystem Assessment. (2005). *Ecosystems & Human Well-being: Synthesis.* Island Press: Washington DC.
60. Everard, M., Orme, J., Longhurst, J.W.S., Grant, M. and Staddon, C. (2014). Healthy people, healthy planet. *The Environment,* March 2014 (Vol.19), pp. 24–29.
61. Department of Health. (2012). *Public Health Outcomes Framework 2013 to 2016.* Department of Health.
62. Antonovsky, A. (1979). *Health, Stress and Coping.* San Francisco: Jossey-Bass Publishers.
63. Wilson, E.O. and Kellert, S.R. (1993). *The Biophilia Hypothesis.* Shearwater Books, ISBN 1-55963-148-1.
64. Wilson, E.O. (1998). *Consilience: The Unity of Knowledge.* Knopf, ISBN 0-679-45077-7.
65. WHO. (2005). *Ecosystems and human well-being: health synthesis—A report of the Millennium Ecosystem Assessment.* World Health Organization, Geneva.
66. World Commission on Dams. (2000). *Dams and Development: A New Framework for Better Decision-making.* Earthscan, London.

67. Everard, M. and Kataria, G. (2010). *The proposed Pancheshwar Dam, India/Nepal: A preliminary ecosystem services assessment of likely outcomes. An IES research report.* The Institution of Environmental Sciences, London. (https://www.the-ies.org/resources/proposed-pancheshwar-dam, accessed 23 April 2015).
68. Everard, M. (2013). *The Hydropolitics of Dams: Engineering or Ecosystems?* Zed Books, London.
69. Palmer, M. and Finlay, V. (2003). *Faith in Conservation: New Approaches to Religions and the Environment.* The World Bank, Washington.
70. Franciscus. (2015). *Encyclical Letter Laudato Si' of the Holy Father Francis on Care for Our Common Home.* Vatican. (http://w2.vatican.va/content/francesco/en/encyclicals/documents/papa-francesco_20150524_enciclica-laudato-si.html, accessed 15 October 2015).
71. Scott, A., Carter, C., Hölzinger, O., Everard, M., Rafaelli, D., Hardman, M., Baker, J., Glass, J., Leach, K., Wakeford, R., Reed, M., Grace, M., Sunderland, T., Waters, R.D., Corstanje, R., Glass, R., Grayson, N., Harris, J and Taft, A. (2014). *UK NEAFO Work Package 10: Tools, application, benefits and linkages for ecosystem services.* UK National Ecosystem Assessment Follow-on programme. (http://uknea.unep-wcmc.org/LinkClick.aspx?fileticket=DYBPt9wHeYA%3d&tabid=82, accessed 18 May 2015).
72. United Nations. (2015). *News on Millennium Development Goals.* (http://www.un.org/millenniumgoals/, accessed 19 September 2015).
73. United Nations. (2015). *Sustainable Development 2015.* (http://www.sustainabledevelopment2015.org/, accessed 19 September 2015).
74. Hoff, H. (2011). *Understanding the Nexus.* Background Paper for the Bonn2011 Conference: The Water, Energy and Food Security Nexus. Stockholm Environment Institute, Stockholm. (http://www.water-energy-food.org/documents/understanding_the_nexus.pdf, accessed 10 September 2014).
75. World Business Council for Sustainable Development. (2009). *Water, Energy and Climate Change: A contribution from the business community.* World Business Council for Sustainable Development. (http://www.unwater.org/downloads/WaterEnergyandClimate Change.pdf, accessed 10 September 2014).

76. UN Water. (2014). *Water, food and energy nexus.* (http://www. unwater.org/topics/water-food-and-energy-nexus/en/, accessed 11 September 2014).
77. Serageldin, I. (2009). Water: conflicts set to arise within as well as between states. *Nature,* 459, p. 163.
78. The Economist. (2011). Unquenchable thirst. *The Economist,* 19 November 2011. (http://www.economist.com/node/21538687, accessed 15 October 2015).
79. United Nations Department of Public Information. (2006). *Ten Stories the World Should Hear More About: From Water Wars to Bridges of Cooperation—Exploring the Peace-Building Potential of a Shared Resource.* (http://www.un.org/Pubs/chronicle/2006/ issue2/0206p54.htm#Water, accessed 15 October 2015).
80. Everard, M. (2013). *The Hydropolitics of Dams: Engineering or Ecosystems?* Zed Books, London.
81. DCDC. (2010). *Global Strategic Trends out to 2040, 4th edition. Development, Concepts and Doctrine Centre (DCDC).* (https:// www.gov.uk/government/publications/dcdc-global-strategic-trends-programme-global-strategic-trends-out-to-2040, accessed 15 October 2015).
82. Everard, M. (2015). *Breathing Space: The Natural and Unnatural History of Air.* Zed Books, London.
83. UNECE. (1979). Convention on Long-range Transboundary Air Pollution. (http://www.unece.org/env/lrtap, accessed 21 June 2014).
84. UNEP. (undated). *Ozone Secretariat.* (http://ozone.unep.org/, accessed 29 December 2013).
85. The Ozone Hole. (Undated). *The Ozone Hole.* (http://www. theozonehole.com/montreal.htm, accessed 14 November 2013).
86. Ramsar Convention. (Undated). *About Ramsar.* (www.ramsar.org, accessed 12 December 2013).
87. IPCC. (Undated). *Intergovernmental Panel on Climate Change.* (www.ipcc.ch, accessed 29 December 2013).
88. IPCC. (2006). *Principles governing IPCC work,* 28 April 2006. Intergovernmental Panel on Climate Change. (http://www.ipcc. ch/pdf/ipcc-principles/ipcc-principles.pdf, accessed 15 November 2013).
89. UNFCCC. (Undated). *UN Climate Change Newsroom.* (www. unfccc.int, accessed 14 July 2015).

90. CBD. (Undated). Convention on Biological Diversity. (https://www.cbd.int/, accessed 18 May 2015).

91. Louv, R. (2006). *Last Child in the Woods: Saving Our Children from Nature-Deficit Disorder*. Algonquin Books, Chapel Hill.

92. Moss, S. (2012). *Natural Childhood*. National Trust, Swindon. (http://www.nationaltrust.org.uk/document-1355766991839/, accessed 15 October 2015).

93. Everard, M. (2011). *Common Ground: The Sharing of Land and Landscapes for Sustainability*. Zed Books, London.

94. Everard, M., Dick, J., Kendall, H., Smith, R.I., Slee, R.W, Couldrick, L., Scott, M. and McDonald, C. (2014). Improving coherence of ecosystem service provision between scales. *Ecosystem Services*, 9, pp. 66–74.

95. Brown, I. and Everard, M. (2015). A working typology of response options to manage environmental change and their scope for complementarity using an Ecosystem Approach. *Environmental Science and Policy*, 52, pp. 61–73.

96. Brown, I., Berry, P., Everard, M., Firbank, L., Harrison, P.A., Lundy, L., Quine, C., Rowan, J., Wade, R. and Watts, K. (2015). Identifying robust response options to manage environmental change using an Ecosystem Approach: A stress-testing case study for the UK. *Environmental Science and Policy*, 52, pp. 74–88.

97. Klein, N. (2014). *This Changes Everything: Capitalism vs the Climate*. Simon and Schuster, London.

98. Hawken, P. (1993). *The Ecology of Commerce*. HarperCollins.

99. Porritt J. (2005). *Capitalism as if the World Matters*. Earthscan: London.

100. Millennium Ecosystem Assessment. (2005). *Ecosystems & Human Well-being: Synthesis*. Island Press: Washington DC.

CHAPTER 7

Co-creating the Symbiocene

Abstract 'Co-creating the Symbiocene' recognises that human pressures will continue to exert significant influence on global ecosystems, much as they have already shaped the Anthropocene. A more comprehensive set of 'artificial selection' criteria is required to direct human decision-making on an increasingly symbiotic basis with the natural processes and 'natural selection' forces that shaped natural systems throughout the Holocene. This would constitute a new synergistic and sustainable epoch—the Symbiocene—with human pressures increasingly sympathetic with natural processes and capacities. A framework for decision-making to achieve this is presented, with worked examples spanning policy areas. All of us exert unique influences through our day-to-day choices and actions, be they wilful or inadvertent, all of which shape the kind of future we are co-creating.

Keywords Symbiocene • Anthropocene • Holocene • Revolution • Future • Symbiosis • Guidance • Decision support

Much of the exploration of humanity's journey to the present in this book has characterised it, to a substantial degree, through successive phases of innovation and of exploitation of water, soil, mined, biological, and energetic resources. Undoubtedly, humanity has progressed to a dramatic degree, from stone tools to spacecraft, subsistence hunter-gathering to

© The Editor(s) (if applicable) and The Author(s) 2016   139
M. Everard, *The Ecosystems Revolution*,
DOI 10.1007/978-3-319-31658-1_7

globe-spanning commodity trading, herbalism to synthesis of sophisti-cated drugs, and energy from muscle power or the burning of biomass to nuclear fission and the direct capture of energy from sunlight. But, as we know, a focus on innovation and exploitation alone without foresight and accountability for wider ramifications is fraught with the kinds of unin-tended downstream consequences we increasingly grapple with today.

Epochs of the Ecosystems Revolution

Recognisably modern humans appeared between 195,000 and 100,000 years ago. Yet the sum total of human tenure on Earth is but a blink of the eye in the greater evolutionary journey of the planetary ecosystem with which we co-evolved. Based on the round figure of 100,000 years, humanity is around 1/45,000 of the age of the planet. Expressed in linear terms, this small fraction would be represented by the readily walked dis-tance between Buckingham Palace and Charing Cross Station in propor-tion to a journey from New York to London, or considerably less than the span of the Brooklyn Bridge in the reverse direction. We are indeed recent arrivals, young and naïve in the face of nature's deeply co-evolved natural intelligence, efficiency, sophistication, and adaptability.

Yet, despite being such planetary late-comers, the impact of humanity upon the biosphere has been both extensive and profound. As discussed right at the start of this book in Chap. 2, the cumulative demands of evolv-ing humanity have kicked biospheric evolution from the Holocene into a new geological epoch defined as the Anthropocene, in which humans rather than natural forces alone have become a dominant influence on Earth's ecosystems. These influences are by substantial majority not benign, including a rapid and accelerating loss of biodiversity, reversal of seques-tration processes remobilising lithospheric chemicals into the biosphere with diverse polluting outcomes including climatic instability, biospheric accumulation of radioactive and synthetic substances, and depletion and destabilisation of ecosystems and their associated functions thereby reduc-ing their services and the resilience of both natural and human systems.

It is my strong contention that future human evolution across and between all sectoral and geographic divides needs to become both wilfully directed and systemically informed—a revolution by choice rather than by chance—by necessity co-creating an increasingly symbiotic relation-ship with global ecosystems across all spheres of societal decision-making. Without this transition, the trajectory of change defining the Anthropocene is likely only to result in increasingly impoverished and conflicted lives as ecosystem capacity shrinks under short-sighted consumption pressures. The need for cross-sectoral and transnational collaboration, not merely around such discrete if imperfect responses as the Intergovernmental Panel on Climate Change (IPCC) or the Montreal Protocol but progressively across all spheres of our activities, must become a unifying force in policy deliberation at all scales, including in product and process innovation. Whilst this may appear initially an impossible and utopian dream, it is in reality far from an idle flight of fancy. Rather, it is a manifesto of what is necessary to secure a decent future for humanity, underwritten by healthy and regenerating ecosystems. And, as we saw in considering the Apollo mission, the Human Genome Project, shifts in practice in the European PVC industry and a range of other guided revolutions, the impossible becomes feasible when a clear and consensual goal drives innovation and collaboration.

Whatever our choice, the future will inevitably be shaped by the human footprint. It is so for all species, interdependent as they are with biospheric processes. However, given our numbers and technological capabilities, the human footprint will inevitably remain significant, whichever of the diverging paths we chose, and so will carry characteristics of the Anthropocene. However, if we are collectively to secure a future of greater opportunity built on securing stable and recovering ecosystems, we will also restore important facets of the Holocene epoch in which the biosphere is defined by natural forces. The future we must necessarily co-create is a synthesis of these two characters: natural forces underwriting our capacity for continuance with human forces driving innovations and a model of progress sympathetic with and ideally restoring the carrying capacity of ecosystems, the many values of which will become integral within societal calculations.

This symbiotic future—humankind rediscovering and making continuing progress with its integral place in the complex biosphere with which we co-evolved as a result of wilful and systemic co-creation—will be an epoch integrating human with natural influences. It is one that deserves a new name reflecting these increasingly symbiotic natural and human forces: the Symbiocene.

EVOLUTIONARY STEPS IN THE ECOSYSTEMS REVOLUTION

As the familiar saying goes, '*A journey of a thousand miles begins with a single step*'.[1] And, as we have seen, a revolution builds from a sequence of evolutionary steps, whether fortuitously or directed. Just like the unanticipated longer-term ramifications of innovation of integrated circuits and of the internet, the consequences of the *Eyjafjallajökull* ash cloud, and the stroke of the metaphorical butterfly's wing upon the inherently chaotic global climatic system,[2] our daily decisions and actions, even the seemingly insignificant ones, matter in terms of the influence they bring to bear upon progress towards a future that will be, aspirationally, sustainable.

All of our decisions have unpredictable end points. However, in some areas of life, our decision-making has a more direct influence in shaping that future. In these areas, three really useful concepts that are gaining wide acceptance in dialogue about systemic interventions—'environmental services', 'anchor services', and 'systemic solutions'—are useful in framing the way we think, and informing the factors and other people we might bring to bear in decision-making.

The concept of 'environmental services' recognises that ecosystem services do not arrive individually from ecosystems, nor are influenced in isolation from other services by management interventions. Rather, ecosystem services are produced as closely interlinked clusters, or 'environmental services'.[3]

However, in practical day-to-day management, there is usually a central driving need or obligation—flood risk management, pollution control, amenity provision, resource security, and so on—framed around a particular ecosystem service. Historically, we have tended to make rather blinkered choices with respect to narrow disciplinary outcomes around that focal service, be it food production, energy generation, or other. However, this central driver can also be regarded as an 'anchor service', around which systemic considerations can be applied to optimise net outcomes across a spectrum of ecosystem services.[4] At worst, this averts unintended negative outcomes; at best, it provides a cost-efficient and sustainable means to pool disciplinary interests and associated budgets to innovate 'systemic solutions'.

As already introduced in Chap. 5, 'systemic solutions' are *'low-input technologies using natural processes to optimise benefits across the spectrum of ecosystem services and their beneficiaries'*.[5] Optimisation of benefits for multiple simultaneous ecosystem services and their associated beneficiaries can contribute significantly to sustainable development by averting unintended negative impacts, increasing ecosystem functioning and resilience, delivering outcomes for multiple service beneficiaries on a more equitable basis, and thereby increasing net economic value. And, as also discussed previously, examples of 'systemic solutions' range from urban green infrastructure to multi-service wetlands including Integrated Constructed Wetlands (ICWs), catchment-based natural flood management, and water resource protection measures, right up to whole-landscape restoration and reanimation. These multiple benefits can be part of a package of 'environmental services', optimised through co-beneficial and potentially co-funded solutions initiated by the need to address a particular 'anchor service'.

Collectively, these three concepts—'environmental services', 'anchor services', and 'systemic solutions'—can help link up policy interests, seeking optimal societal benefits founded on an ecosystem-centred approach. Putting these concepts into action might follow a number of steps that break with historically narrow policy formulation and decision-making practices. A practical framework identifying questions that can help direct decisions towards systemic outcomes, relevant across societal sectors and interests, is outlined in Fig. 7.1.

How the Framework May Expand Decision-Making

The following paragraphs in this sub-section each address an everyday challenge and how it can be addressed using the processes articulated in Fig. 7.1.

Urban Design

In urban design, this may manifest as: clear definition of needs in a development scheme (e.g., bring more ecosystem services benefits into a piece of urban planning); identification of how ecosystems already contribute to the viability of that development (such as natural drainage lines, air-cleansing trees, and 'blue' networks); exploration of how different urban design options can achieve co-benefits, or could conflict, with desired outcomes across a range of linked policy areas; identification of options for nature-based urban design that may optimise outcomes across multiple ecosystem services; determination of where departmental budgets may be pooled to optimise identified potential co-benefits (e.g., where a sustainable drainage design can be adapted also to provide amenity, biodiversity, and other urban co-benefits); recognising shifts in the policy environment that may be more favourable to sustainable urban design, for example, by promoting net societal value rather than lowest capital costs as the central

What is the problem or management need?

What natural processes already provide services that help manage the problem or need, or could be used to do so?
- What ecosystem services already, or could potentially, support the desired outcome?
- Can nature-based solutions, or working with natural processes, help address the problem or need?
- What is the 'anchor service' around which a more system solution may be considered?

What are the ramifications of proposed solutions across the spectrum of ecosystem services, including both formerly unforeseen impacts and co-benefits?
- Solutions to problems or needs might historically have been met in narrow disciplinary terms, but it is important to assess the breadth of impacts for ecosystem services (for example using the IES *Ecosystem Services Assessment* guide[6]) and their beneficiaries if equitable, sustainable and potentially economically efficient outcomes are sought.
- How might solutions be modified to cause less unintended negative outcomes whilst optimising net benefits across services?

Who are the beneficiaries of both potentially positively and negatively affected ecosystem services, who are their representatives, and how can they be brought into a stakeholder-based process of optimising ecosystem service outcomes?
- Who is affected through ecosystem service impacts by potential management options?
- Capturing the multiple types of knowledge held by multiple affected stakeholders is important to broaden the basis of decision-making. Societal/stakeholder participation in decisions taken at appropriate scales can ensure that all service beneficiary (or victim) perspectives are factored into decision-making.

What 'systemic solutions' –"…*low-input technologies using natural processes to optimise benefits across the spectrum of ecosystem services and their beneficiaries*" – can deliver greatest net societal value?
- Taking account of ramifications for all services and their beneficiaries, what solutions – ideally making use of natural processes with their associated low inputs – offer the greatest opportunities for optimisation of benefits and avoidance of negative externalities?

Can resources (financial, influence, labour, etc.) be pooled to achieve greater net benefits and cost savings?
- Can linked interests, associated budgets, labour and other forms of support be pooled to optimise cross-service benefits and hence net value across society?

What policy instruments can be used to promote optimal and desired outcomes?
- How can the policy environment in its broadest sense – formal and informal including, for example, legislation, taxes and subsidies, market-based instruments, common law, assumptions built into decision-support models, by-laws and social protocols – be marshalled to achieve optimal outcomes?
- Where are the enablers and what are the barriers to systemic outcomes?

Learning
- What learning emerges from this process – for example about stakeholder working, assessment of net benefits, policy reform, communication, etc. – that can inform more systemic practice in future?

Fig. 7.1 Conceptual framework for directing decisions towards systemic, multi-service outcomes

presumption in urban planning decision-making; and resultant lessons for use in future development strategies, which may include instituting a process of cross-service benefit-cost assessment for development schemes taking account of all ecosystem services and their beneficiaries.

International Relations

In international relations, use of the framework may manifest as: clear definition of the issue to be resolved (e.g., co-management of a transboundary river); identification of ecosystem services upstream supporting flows and maintaining water quality; consideration of the various outcomes of different catchment management options across a range of ecosystem services; cross-border agreement of best management options to optimise ecosystem service outcomes that may increase overall ecosystem capacity to the benefit of neighbouring riparian states; financial agreements potentially with international subsidies that reward land uses protective of water quantity and quality advantageous to both states; application of policy instruments cementing co-beneficial outcomes; and lesson-learning to assist in future cross-border natural resource-sharing arrangements.

Air Quality Management

In air quality management, the perceived problem may be a local exceedance of air quality pollution levels in a town; rather than simply diverting traffic and the problem elsewhere, green infrastructure might promote more sustainable transport (walking, cycling, etc.) through the town whilst also settling particulate matter from the air; other potential co-benefits that can be designed into green infrastructure include provision of natural flood management, spaces for amenity and healthy exercise, visual and sound buffering and wildlife corridors; bringing together relevant departments and external stakeholders can seek to optimise benefits in planning; the green infrastructure solution is inherently a 'systemic solution' using natural processes to optimise ecosystem service benefits; as such, it may also serve as a basis for co-funding between departments and other interests, with multi-benefit solutions delivering optimal outcomes for lower overall investment; sustainable development duties and strategies can be used to promote this integrated, 'best value' approach; however, lessons may be learned, for example, about whether narrowly framed legislation

and funding conditions for some discipline-bound issues may have to be
challenged where they restrict the realisation of cross-disciplinary benefit.

River Restoration

In river restoration, the challenge may be to improve flood regulation
and the net value of an urban stream to local people (as in the case of the
River Quaggy and Mayes Brook in London as addressed in Chap. 5); this
may entail 're-naturalising' the river through floodplain reconnection to
restore aspects of its functioning; this has the potential to enhance a range
of linked ecosystem services from provision of educational to recreational
resources, regulation of flood and microclimate, enhancement of biodiversity and aesthetics, and regeneration of the adjacent neighbourhood; the
significant multi-beneficiary potential of such a restoration scheme means
that planning is most constructively undertaken through engagement with
multiple affected stakeholders; the rehabilitation of ecosystem functioning
and benefits is itself a 'systemic solution' addressing multiple beneficial
outcomes through natural, low-input processes; this linkage across disci-
plines means that there are significant opportunities to pool and increase
net value from disparate funding streams, including neighbourhood support; as for the air quality management example above, sustainable development duties and strategies can be used to promote this integrated, 'best
value' approach; also, narrowly framed legislation and funding conditions
for some discipline-bound issues may have to be challenged where their
reductive framing potentially restricts the realisation of greater cumulative
cross-disciplinary benefits.

Flood Risk Management

In flood risk management: the primary driving management need is to
reduce risk to built and farmed assets; catchment dynamics already deliver
natural flood regulation services and could be enhanced; where a 'natural flood management' approach is adopted (as described in Chap. 5),
this can also regenerate fishery recruitment, enhance biodiversity, provide
income to rural land owners paid for water storage, sequester carbon,
cycle nutrients, and so on; identifying and engaging these co-beneficiaries
can support planning for integrated outcomes of greatest net cross-service
benefit; this approach is inherently a 'systemic solution', making use of
natural processes to optimise outcomes across a range of services and their

beneficiaries; it thereby provides opportunities to pool effort and investment such as better targeting of agri-environment expenditure and also budgets and civil society involvement in fishery and nature conservation; 'best value' judgements can favour this approach, although legacy regulations and subsidy schemes may be narrowly 'ring-fenced' requiring innovation if they are to be redirected at pooled net beneficial outcomes; which are amongst the most significant learning points for out-scaling to other flood risk management schemes and to guide policy reform.

Management of Road Verges

In considering the mowing of banks on trunk roads, the management need is mainly today driven by road safety considerations; seasonal dieback of vegetation manages the risk in winter, so the main concerns are summer growth and succession to bramble and scrub that may reduce road safety; however, it is becoming increasingly clear that motorway and other major road verges are important wildlife corridors including serving as valuable reservoirs of plant species and associated insects and other wildlife that are adversely affected if cutting occurs before flowers have set seed[7] as well as serving a potential role in natural flood management; clearly then, more ecosystem services and associated interests are affected by road verge mowing than has historically been accounted for, and therefore these interests need to be brought into the planning process for a more sustainable management regime; a 'systemic solutions' approach would explore how natural, benefit-yielding processes can be considered to optimise benefits across a range of services; collaboration across departments and social sectors, including for example drainage engineers and nature conservation NGOs and academics, would help inform a more multifactorial approach to verge maintenance; these could be argued for on the basis of sustainable development duties and biodiversity strategies; however, there will almost certainly be resistance from legacy, narrowly framed legislation and budgets that will need to be overcome through reform of agreed operational protocols if 'nature unfriendly' outcomes are not to be perpetuated.

Product Innovation

In product innovation, following the proposed framework may take the form of: clear definition of need as opposed merely of innovation for its own sake (such as a new, more sustainable office chair); how ecosystems may already support this need (sources of wood, energy, etc.); how chair design can result in the least impact on natural resources, or potentially be founded in restorative schemes such as FSC wood; co-benefits from a nature-based approach (such as generation of improved or additional resource stewardship schemes or in recycling infrastructure to recover value from end-of-life products); collaboration with other businesses and societal sectors (such as waste handlers and regulators important in enabling recycling) that may want to share investment in co-beneficial schemes; identification of favourable policy instruments and those that might need reform to favour sustainable business practices; and review of principal learning to improve future design processes.

Pollination and Pest Control in Orchards

In addressing linked pollination and pest control in orchards, primary needs are to avert plant diseases without excessive use of chemicals threatening pollinating insects or human health; pollinators and predators of pest organisms occur naturally; careful management of adjacent hedgerows and other important habitats can protect and enhance populations of both pollinators and pest-consuming predators such as wasps and insect-eating birds; co-benefits from this approach include improved prospects for a range of other farmland wildlife and landscape aesthetics; co-beneficiaries include those sharing these interests; a 'systemic solution' preferable to higher inputs of potentially problematic chemicals is to enhance adjacent ecosystems naturally providing these linked beneficial services; if a consortium of common interest can be assembled, collective labour and funding can bring about collective cross-service benefits; this may break with a traditional focus on single outcomes, but is not only feasible but of greater value to all parties; and lessons achieved en route can inform other areas of pest and pollinator management and policy.

Home Energy

In home energy planning, primary policy driving forces relate to the linked goals of carbon reduction and energy efficiency. Naturally occurring energy—in sunlight and in the air and ground heat as well as in wind and water currents—not only represents a potentially renewable source but also already provides some services such as solar gain and natural cooling. Home-scale energy solutions can be hydrologically neutral, for example, roof-mounted solar panels, alleviating environmental pressures arising from centralised generation technologies; co-benefits may be achieved by upscaling to community-based generation, including techniques that provide co-benefits for nature such as habitat within solar farms. Co-beneficiaries from promotion of home- and community-scale energy technologies are primarily global through reduced pressure on the climate; whilst not fully 'systemic solutions', there is at least some provision for co-benefits across ecosystem services through the use of natural energy flows. Pooling of resources occurs through supportive measures in the policy environment, such as FIT, with government taking up the role of public beneficiaries; this kind of positive promotion, as well as experiences of local authorities seeking to block installation of novel technologies largely on the basis of not knowing about the relevant legislation (as experienced when the author installed photovoltaic panels), can inform areas for education and progressive policy reform to promote societally beneficial outcomes that are affordable for the homeowner.

Regional Strategy

In regional strategy, the example of the South African Aerotropolis has already been outlined in the preceding Chap. 6, full of well-informed systemic intention but thwarted in delivery due to traditional commercial short-termism and weak enforcement. However, another practical and more successful example of this integrated approach in practice with which I have been involved is the development of a wetland strategy for the greater city region of Colombo, Sri Lanka.[8] The principal driver of this World Bank-funded project was to safeguard flood regulation and other linked services as well as the economic value of the network of wetlands throughout the region in the face of rapid urbanisation and changes from rural to urban livelihoods. The project was challenging, entailing surveys of the physical extent of diverse wetland types—canals, lagoons, marshes, and abandoned and active paddy, all with and without encroachment by largely informal development—with extensive assessments of baseline multi-taxa biodiversity, hydrology, water chemistry, ecosystem services, and socio-economic contexts. The alchemy of the research project occurred in the synthesis of these diverse attributes to determine how and where ecosystem service benefits were produced and expressed across the interconnected wetland network and its close intermingling with sprawling urban infrastructure. This in turn resulted in valuation of wetlands, not in an isolated sense of a fixed price on each site but in terms of the value of their interconnections. Multiple services recognised as stemming from this diverse and extensive wetland resource included, amongst others, flood regulation, drought storage, physico-chemical water purification, carbon sequestration, microclimate amelioration, provision of fish and other food, ornamental and medicinal goods for both formal and informal use and trade, aesthetic and traditional landscapes and uses, recreation and amenity opportunities, soil formation, and habitat for wildlife. Part of the study included seeking novel economic uses for wetlands such as abandoned paddy, the historic marketed value of which was in decline, yet for which it was vital to retain important non-marketed 'environmental services' such as regulation of cross-city flooding, microclimate, and habitat for wildlife. The Colombo wetland strategy thereby connects with and seeks linked benefits across multiple policy areas, also seeking to pool budgets for realisation of those benefits not merely through novel economic opportunities but also the linking up of financial and political capital across co-beneficiary policy areas, including education, security, heritage, and flood risk management, as a basis for more cost-effective and integrated wetland protection and use.

TOMORROW'S WORLD

We face a long and demanding journey from where we stand today towards the grail of fully systemic thinking and practice, within which society holds the integrated services provided by ecosystems equally as axiomatic as economic and social priorities. For all the good examples, it is the mainstream of practice that must morph into something that provides humanity with positive prospects for the future, as well as mitigating risks and liabilities in the present. Without this paradigm flip, science and logic tell us that we continue to undermine the most fundamental resources upon which future security and wellbeing depend.

Tomorrow will happen, for the worse as an extended declining Anthropocene or because we have seriously engaged with the threats and the promise to co-create a Symbiocene of unfolding opportunity supported by a secure and recovering biosphere. The nature of that tomorrow lies in the hands of today's global stewards. So who are the principal stewards of changes necessarily constituting an ecosystems revolution?

Despite the environmental grounding of this ecosystems revolution, consequences for 'the environment' are far too important to be left to environmentalists alone. Indeed, if the broader environmental community, howsoever defined, is ultimately seen as sole 'owner' of the Ecosystem Approach then all is lost as this would serve only to continue to marginalise a transition that must necessarily span and engage all sectors of society.

We have also seen that politics generally follows rather than leads, so waiting for meaningful leadership from this quarter is a vain hope. That said, some of the stewards of tomorrow may be political leaders in different government strata who place strategic progress before self-gain, the perpetuation of yesterday's fixed assumptions, the pressures of vested interests, and short-term appeal to the electorate. Other influential stewards may be thought leaders in universities or activists in NGOs. But ultimately, the ecosystems revolution is about systemic change. Leadership then is not 'owned' by any one sector, nor should any sector defer to others in taking it.

Senior colleagues and I have debated for some years when drafting some government strategy documents whether we are bold enough, or when the world will be adequately sensitised, for us to drop the 'eco-' prefix when presenting 'ecosystems' and the 'Ecosystem Approach'. The mere mention of 'eco-anything' suggests to many that this is all to be delegated to the 'greenies', perpetuating the notion that ecosystems and their services are not central to the interests of all and that concern about them can be delegated to others once the big financial decisions have been made. The reality is that the Ecosystem Approach, defined by its 12 'complementary and interlinked' principles spanning societal choice and inclusion, natural functions and limits, economic context, governance and the wider consequences of decision is about engaging all sectors into consideration of broader contexts and consequences. We should therefore really be talking about a 'Systems Approach', in which it is understood that supporting ecosystems stand as axiomatic alongside economic, social, and equitable considerations. Is society ready yet to realise that 'ecosystems' are such a fundamental concern, and not merely an altruistic afterthought? Perhaps not yet, but this is an interim goal against which to target our innovation and efforts.

The journey ahead is also challenging for us all, including for long-term proponents of sustainable development who also have to recognise that their views must evolve to accommodate a new paradigm. For example, it has become abundantly clear that our former approach to sustainable development, mainly focused on lightening and eventually seeking neutrality in our 'tread on nature', is wholly lacking in ambition and likely positive outcome given massively declining global trends in the ecosystems vital for securing our future. We have to elevate our vision and ambitions to rebuild lost and degraded ecosystem capacities if we are to enjoy a future defined by anything other than declining opportunity and realisation of human potential.

Recognising the long-term nature and the profound challenge of achieving the ecosystems revolution, we have also to take encouragement from the fact that the journey has been embarked for more than a century. Not only do we need all the reassurance we can get, but we can also draw from former successes important lessons to take forward as evidence in the drive progressively to link up all societal sectors in a conjoined goal.

Furthermore, we all also have to realise that we are more powerful than we generally assume, particularly in our influence upon and wider interactions with the rest of humanity. Like the metaphorical insect's wing in the 'butterfly effect', or indeed the apocryphal janitor who was helping put a man on the moon by keeping dust out of the spacecraft's workings, the influence we can all bring to bear based on our unique perspectives and spheres of influence can have unforeseen 'knock-on' effects for the whole complex and chaotic societal system. We have, as Niels Bohr and Albert Einstein remind us (see Chap. 4), no idea just how far our influence may resonate, even our ostensibly trivial and inconsequential words and actions. All we can say for sure is that our small interventions will have at least some consequence, so are always worth making as a unique personal contribution to the broader and visionary ecosystems revolution.

You and I, right now, are shaping tomorrow's world in all the things we do, as indeed in the things that we do not do, fashioned as our actions and inactions are by the intent behind them. And so it is to you and your unique views and contributions, and all the other billions out there with whom we share this marvellous planet, that this book is dedicated.

Let us use our gifts and opportunities well, with clear and far-sighted intent and with systemic consideration to co-create a Symbiocene of secure prospects and expanding opportunity for all.

NOTES

1. Literally, '*A journey of a thousand miles starts beneath one's feet*', a famous Chinese proverb, quoted in the Tao Te Ching and ascribed to the ancient Chinese poet and philosopher Laozi.

2. In chaos theory, the butterfly effect is a metaphor used to exemplify the sensitive dependence on initial, even apparently inconsequential conditions of outcomes in nonlinear systems, the flapping of the wings of a distant butterfly several weeks previously potentially influencing the formation, timing, and path of a hurricane on the other side of the planet.

3. Schomers, S. and Matzdorf, B. (2013). Payments for ecosystem services: A review and comparison of developing and industrialized countries. *Ecosystem Services*, 6, pp. 16–30.

4. Everard, M. (2014). Nature's marketplace. *The Environmentalist*, March 2014, pp. 21–23.

5. Everard, M. and McInnes, R.J. (2013). Systemic solutions for multi-benefit water and environmental management. *Science of the Total Environment*. 461–462, pp. 170–179.

6. Everard, M and Waters, R.D. (2013). Ecosystem services assessment: How to do one in practice. Institution of Environmental Sciences, London. (https://www.the-ies.org/sites/default/files/reports/ecosystem_services.pdf, accessed 1st March 2016.)

7. Clover, C. (2002). Motorway verges to be wildlife havens. *The Telegraph*, 20 March 2002. (http://www.telegraph.co.uk/news/uknews/1388271/Motorway-verges-to-be-wildlife-havens.html, accessed 17 February 2015).

8. McInnes, R.J., Jaksic, S., Everard, M. Bandara, A., Amaralal, L., Weerakoon, D., Moynot, G., Salmon, G., Bargier, N., Piyadasa, R., Rigaudiere, P. and Ranwala, A. (2016) *Metro Colombo wetland management strategy*. Unpublished report to Sri Lanka Land Reclamation Development Corporation and World Bank. Signes: Paris, France. 107 pp.

INDEX

Printed in the United States
By Bookmasters